3 4028 02839 8783

W9-AQZ-617

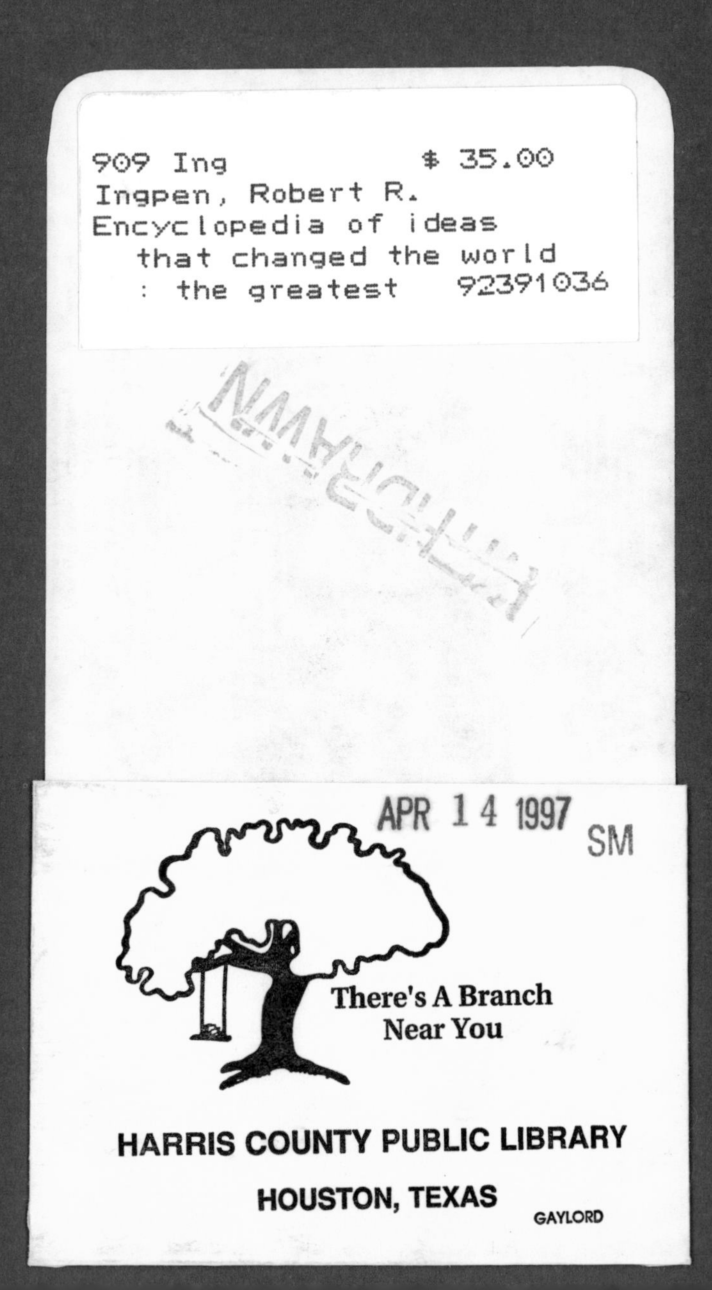
909 Ing $ 35.00
Ingpen, Robert R.
Encyclopedia of ideas
that changed the world
: the greatest 92391036
WITHDRAWN
APR 14 1997
SM
There's A Branch
Near You
HARRIS COUNTY PUBLIC LIBRARY
HOUSTON, TEXAS
GAYLORD

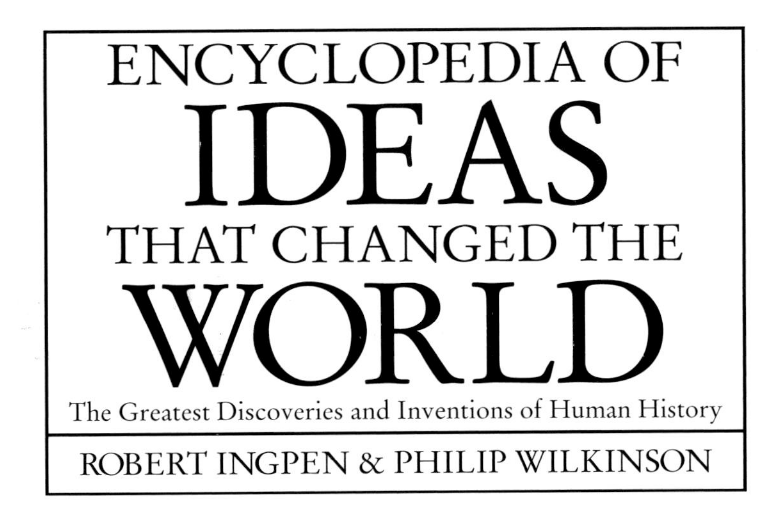

ENCYCLOPEDIA OF IDEAS THAT CHANGED THE WORLD

The Greatest Discoveries and Inventions of Human History

ROBERT INGPEN & PHILIP WILKINSON

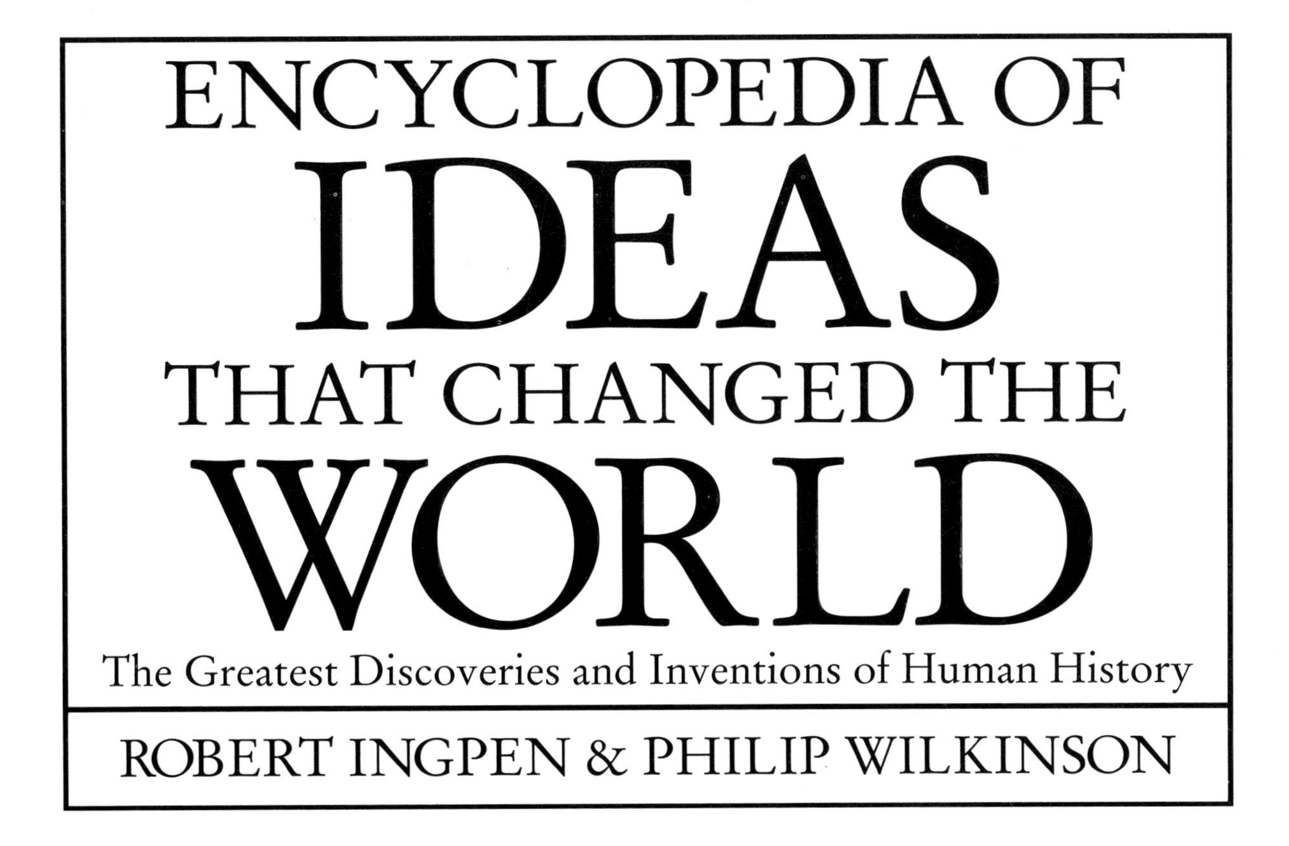

ENCYCLOPEDIA OF IDEAS THAT CHANGED THE WORLD

The Greatest Discoveries and Inventions of Human History

ROBERT INGPEN & PHILIP WILKINSON

Acknowledgements
The author would like to thank Lawrence A. Bornstein, Terry Sharrer and Teresa Briscoe for reading the text and making many valuable comments, and Zoë Brooks for her unstinting support.

ENCYCLOPEDIA OF IDEAS THAT CHANGED THE WORLD

VIKING STUDIO BOOKS
Published by the Penguin Group
Penguin Books USA Inc., 375 Hudson Street, New York, New York 10014, U.S.A.
Penguin Books Ltd, 27 Wrights Lane, London W8 5TZ, England
Penguin Books Australia Ltd, Ringwood, Victoria, Australia
Penguin Books Canada Ltd, 10 Alcorn Avenue, Toronto, Ontario, Canada M4V 3B2
Penguin Books (N.Z.) Ltd, 182-190 Wairau Road, Auckland 10, New Zealand

Penguin Books Ltd, Registered Offices: Harmondsworth, Middlesex, England

First American edition
Published in 1993 by Viking Penguin, a division of Penguin Books USA Inc.

10 9 8 7 6 5 4 3 2 1

Produced by Dragon's World Ltd., Limpsfield, Surrey RH8 ODY, England

Designer: Robert Ingpen
Editor: Elizabeth Radford
Design Assistance: Tom Deas
DTP Manager: Keith Bambury
Editorial Director: Pippa Rubinstein

ISBN 0-670-84642-2

CIP data available

Printed in Italy
Set in Garamond

CONTENTS

INTRODUCTION

Imagine a world without anaesthetics or antiseptic surgery: the surgeon thinks twice about performing an operation because it is likely to result in extreme pain; worse still, the patient stands a good chance of dying from infection. When an operation *is* performed, it has to be done with such speed that little precision is possible. The idea of surgery strikes terror into the heart of patient and relatives alike. Life is nasty, brutish and short.

Go back further, to a world where the steam engine and allied inventions had not yet enabled industrialization. Most people live in rural surroundings. They do not 'go to work', as do most people today. They work on the land near their home, and live with their animals – often in the same building. Life is bounded by the seasons, and the rising and setting of the sun.

Then imagine an even earlier world, without the wheel. It is a world in which the only land transport is on horseback or on foot, in which all but the simplest machines are inconceivable, in which large numbers of people are needed to shift even a moderately heavy load, in which long-distance travel is the preserve of the very rich, and thus the very few.

When we think of such pictures from our past, we may have some idea of the impact of mankind's inventiveness. Similarily, many things that today we take for granted were unthinkable a hundred years ago. Convenient air travel, nuclear power, the cure or prevention of a whole range of diseases, the appearance of computers and microprocessors so small that they can be built into everyday domestic appliances: all of these would have been hardly imaginable to people living at the end of the nineteenth century.

An endless search

How did these momentous developments come about? There are as many answers to that question as there are inventions. But there seems to be a fundamental need for people to find better ways of doing things, and thus means by which our lives can be improved. This evolution began when our earliest ancestors started to experiment with stone tools, and will continue as long as there are humans on the Earth, as long as doctors strive to find cures for our diseases, engineers test new materials, and physicists search for new ways of generating power.

This book is about these ideas and the quests and experiments – even the accidents – that led to them. The stories that lie behind the great human inventions are some of the most fascinating in our history. They range from the simple (such as the discovery and exploitation of fire) to the complex (the intricate calculations needed to predict nuclear reactions). They encompass tales of painstaking trial, error and retrial (such as the Wrights' tests of their flying machines), to happy accidents of the laboratory (the discovery of a new plastic). They extend from stories of defiant risk-taking (like Jenner's first vaccinations) to accounts of lengthy, methodical research.

The process of invention

Interesting as such stories are, the closer we look, the more it is clear that their ideas do not come out of a void. An important invention is most often the result of years of dialogue between scientists, and long periods of experiment, trial and error. It also relies on the succession of preceding inventions. Film was impossible without the technology developed for still photography; powered flight was difficult without lightweight engines, and so on.

Furthermore, key technologies appear independently around the world. In the West we think of Gutenberg as the inventor of printing with movable type. But printers in China were using similar technologies for centuries before Gutenberg, and craftsmen in Korea were inventing movable type, independently, at the same time. Other 'Western' inventions, such as paper or gunpowder, are often taken for granted, without their far-eastern origins being remembered. So the relationship between East and West is a theme which recurs in this book.

This state of affairs often makes it extremely difficult to define exactly what was the key invention or 'great idea' that changed the world. Was it, for example, the invention of the internal combustion engine, its application in the first motor car, or the use of mass production in automobile manufacture that changed the face of transport the world over? Was the invention of film, or that of television more influential?

In these examples, all the ideas were important and all happened quite close together in time. But sometimes the story begins centuries before the crucial turning point. Clay pots, for example, must have been made long before someone stumbled on the fact that their usefulness and durability could be increased by firing them at high temperatures. So it is the interaction between the ideas, the interplay between researchers, scientists, manufacturers, and users of their products, that is important. The users, perhaps, above all, since grasping the way in which an idea can be used and can influence events or benefit peoples' lives, are as important as the original idea.

Invention and need

So what is great about the 'great ideas' in this book? Above all, it is the influence they have had on the world, the changes they have wrought on our lives. It is difficult to imagine how we got along without most of these innovations. But this obvious 'need' is not always apparent before the satisfaction of that need.

Sometimes, it is true, there is a need crying out for an invention to fulfil it. This is the imperative felt by most of those who research in the field of human health. Jenner was driven by the desire to eradicate, or at least to reduce, the risk of a deadly disease; Lister by a wish to make the operating theatre a safer place for the patient. A similar obvious need was felt by the mining engineers who pioneered the steam engine: as coal mines went deeper in search of further fuel reserves, so the need to pump out water increased. But few, if any of those engineers would have predicted the wider effects and influences of the engine they helped to bring into being – the coming of the railways, the increase in manufacturing, the transformation of the working lives of millions of people.

In other instances there is a more generalized need. As industry began to expand in the eighteenth and nineteenth centuries, better transport was an obvious goal of those who wanted to cart raw materials around, or to trade in industrial end-products. Such an improved transport network could come about in several different ways – and it did. One way was simply to build more of the existing arteries of transport: more roads, more canals. Another was to improve these by better building techniques, hence the improved roadmaking formulae of men like Telford and Macadam. Yet another was to invent a totally new form of transport: the steam-driven railway.

Often, a need is waiting to be perceived, such as printing with movable type. The circulation of manuscript books, each laboriously handwritten, was no way to transmit ideas in the intellectual ferment of the eve of the Renaissance and the Reformation. When Gutenberg and his co-workers came up with a solution, it was seized upon with surprise and enthusiasm – in a couple of decades there were presses all over Europe, and thousands of printed books were circulating.

But on some occasions the inventor is simply ahead of the need. Many of the pioneers of flight, sending vehicles across the skies, or Marconi, sending invisible messages through the air, must have seemed quite eccentric to their contemporaries. Yet a dogged pursuit of scientific enquiry, and an absorption in what they were doing, kept them going. In the end, the need for their inventions was clear. In the case of some researchers, an outside event speeds the coming of the need. The First World War, for example, stimulated the creation of the tank, although caterpillar-tracked vehicles had been around, waiting for a purpose, for more than twenty-five years, while the Second World War helped to usher in the jet engine and the rocket.

Bright ideas

So we need to be aware of the historical context when considering human inventions. But such an awareness should still not make us forget the power of sheer inspiration, of the 'ideas person', who works alone and in comparative obscurity. This can require both dogged dedication and the ability to think in a quite original way, to have, in short, the equipment of the scientist, but to see beyond conventional scientific boundaries and definitions to something quite new. This can entail 'breaking the rules' to try unlikely ideas. More than one inventor has admitted that if he had known more about the process he was working on, he would not have bothered to try an idea out at all. But even if you do know, you need the courage to try, and the courage to fail. For it takes courage and initiative for people like the Wright brothers, testing their flying machines, or Marconi doing his early work on radio, to stick to the idea, to experiment, adapt, and carry on thinking and working until it is right. People like these have earned their place in our own thoughts, and have changed our world.

TIMELINE OF IDEAS

	AGRICULTURE AND BUILDING	TECHNICAL ADVANCES	USE OF RESOURCES	MATERIALS	COMMUNICATIONS
BC		Pebble tools (2 million+ yrs)			
		Flint hand tools (1.3 million yrs)			
100,000+	Wooden huts, France (125,000 BC)		*Homo erectus* using fire (300,000BC)		
10,000	Farming begins (Japan)				
8000	Agriculture (Asia)				Dugout canoes
6000	Houses of wood and thatch (Europe) Mud-brick buildings (W. Asia)		Ore smelting (W. Asia)		
4000	Stone plough shares (China)			Bronzeworking (W. Asia)	Boats with sails
		Pottery kilns (China)			Writing (Mesopotamia)
2000	Stone construction (Egypt)		Wooden ploughs and plank boats	Iron working (W. Asia)	Wheel (Mesopotamia)
		Trained doctors (Egypt)		Glass (N. Asia) Pottery (Mesopotamia)	
1000	Domes and vaults				Galley depicted
500	Classic hand tools (Assyria)		Lathe developed		
400		Hippocrates' medical writings		Iron casting (China)	Assyrian wood rafts Compass (China)
300	Iron plough shares (China)	Greeks use simple machines	Water wheels (France)		Paper (China)
200	Pantheon dome	Galen's medical writings			Construction of Appian Way
100	Romans use concrete		Watermill first described		
AD 200		Power hammers (China)		Porcelain (China)	
500		Stepladder of water wheels (France)			
1000	Great Stone Bridge (China)	Persian windmills			Printing wood-blocks (China)
1400					Movable type and books in West
1600		Savery's mine pump			
		Jansen's microscope	Coke for smelting		
		Blood circulation described			Lippershey's telescope
				Lead crystal glass	Newcomen's steam engine
1700	Turnips and crop rotation	Tull's seed drill			Tresaguet's formula for road-making
		Watt improves steam engine			
	Iron aqueduct (France) and factories (England)	Whitney cotton gin		Crucible steel	Cugnol's steam carriage
		Jenner's vaccination and Galvani's electricity experiments			Montgolfier brothers' first balloon flight
		Maudslay's lathe			
1800		Stethoscope in use	Volta's electrical experiments		Trevithick's carriage
	McCormick's reaper	Colt patents revolver	Faraday's electrical experiments		Stephenson's locomotive
	'Safe elevator' from Otis	Nasmyth's steam hammer			McAdam roads
		Mass production of sewing machine	Henry's electric bell		Cooke/Wheatstone's telegraph
1850		Antiseptic surgery			De Rochas four-stroke cycle engine
				Bessemer converter	
	Steam plough				Bell patents telephone system
		Carbon filament lamp		Parkesine produced	
					Edison's first sound recording
	Reinforced concrete	First generating stations			
	Barbed wire	Parson's turbo-generator built			Marconi experiments with radio
			Becqueral and the Curies work on radioactivity		
	Eiffel Tower	Electric fires and fans introduced			Daimler's internal combustion engine
1900				Bakelite produced	Wright Brothers' first flight
			Liquid fuel for rockets		
					Jet engine built
		Turing's paper on computers published		Nylon first made on commercial scale	
			Ferris's nuclear pile built		
		V2 missile launched			
1950			Space rockets		

This time-line shows the major 'great ideas' discussed in this book in relation to each other, and highlights the uneven spread of these ideas during the course of human history; indeed, the last 800 years comprise half the chart, which spans over two million years. The last date given is 1950, reflecting the author's feeling that although enormous strides have been taken in all areas of human activity since then, these tend to be developments, rather than completely new concepts.

THE FIRST INVENTIONS

What were the very first inventions? If developing and applying ideas is one of the basic abilities of the human race, answering this question should take us back to our true beginnings. And so it proves: look back to the earliest evidence of humanity, at sites such as Olduvai Gorge in East Africa, and you find simple pebble tools, proof of the ingenuity of our ancestors some 2 million years ago.

Go to other places, such as the caves of hominid *Homo erectus* in Choukoutien, China. Here are charred remains suggesting the use of fire between 1 million and 300,000 years ago. Fire would have been invaluable in the cold northern areas that *Homo erectus* inhabited; not just for heating, but also for cooking, lighting, and even protection from predators.

Visit sites in Europe that some 60,000 years ago were the homes of the Neanderthals. You will find human skeletons that have been buried in a deliberate and careful way, indicating the origins of ritual and ceremony. The bodies were arranged carefully, and grave goods buried with them. This was more than the utilitarian removal of a corpse, it was a ceremonial act, and may even indicate a belief in an afterlife.

And look at other European sites, such as Lascaux in France, or Altamira in Spain. On rock faces, cave walls and pieces of bone, early people carved and painted pictures and patterns. These are early artistic masterpieces, made by painters of the utmost sophistication, before 10,000 BC. *Homo sapiens*, wise man, was also artistic man.

Toolmaking, use of fire, ritual, art: they seem a curious group, especially in a book about the story of technology. But they typify something vital to the history of invention: the coming together of technical skill (craftsmanship, knowledge of materials and the environment) with the imagination (the ability to make conceptual leaps and link different technologies together; the willingness to seek solutions in unlikely places). This fusion is surely one of the vital factors in making us the humans we are.

These four achievements also look forward to many of the other human ideas described in this book. The earliest pebble tools developed into hand-axes and into the specialized implements – knives, saws, awls, chisels – that are the ancestors of modern tools. And the needs they fulfilled also led to the production of a host of machines, from pulleys to robots. The discovery of fire looks forward to the harnessing of so many other natural resources – wind, water, steam, and so on – which today we take for granted. Even art and ritual, while not technologies in the accepted sense of the word, are forms of communication prefiguring many other, more recent innovations in the field of communications. The first inventors, then, were indeed influential, and we are still in their debt today.

EARLY TOOLS

The creation of tools set humans apart from other mammals and increased their abilities both inside and outside the home

Some 4 million years ago, our ancestors started on a course that was to be one of the most important developments in the human story: they began to walk upright. One effect of this was to free their front limbs from the task of locomotion, allowing the hands to become properly hands for the first time. The already strong powers of mind-body co-ordination could develop further, and the hands were free to make things. From this point on, the unique human contribution to the transformation of the environment began.

Finding tools

Near the beginning of this story, the early hominids probably learnt to extend both the length and the ability of their arms by using simple tools. To begin with, this may have been no more than the sort of elementary tool-use that zoologists have noticed in other animals. Apes, particularly chimpanzees, for example, have often been observed using tools. They use lengths of bamboo to increase their reach, and blades of grass to fish for termites in termite mounds.

In a similar way, our ancestors would have begun to pick up sticks and stones to help them perform basic activities. Strong sticks could help them to grub up roots out of the ground; stones could be used as primitive choppers or pestles. It would not be long before someone noticed that some chipped stones had sharp edges and could be used (like knives) to cut up meat or plant food.

No one knows when the early hominids started to use simple 'found' tools in this way. But chipped pebbles that could have been used like this have been found at hominid sites in Africa. They were probably made more than 2 million years ago.

To begin with, this type of activity had its limitations. It appeared, as it does with the chimpanzees, in response to a particular problem: once the problem is solved, the tool is abandoned. There is no foresight or forward planning involved – and only a very limited imagination.

The birth of tool-making

The next important step was the uniquely human contribution. Having observed that a chipped stone had a sharp edge, the early hominids started to chip away at one stone with another, whittling away at the surface of the stone until a razor-like sharpness was obtained. True tool-making had begun.

The first people to do this were probably hominids in East Africa, who lived between 2 and 1.5 million years ago. Two of the best-known sites where such tools have been found are Koobi Fora, east of Lake Turkana and Olduvai Gorge, east of Lake Victoria. Both were occupied by members of a hominid species that many prehistorians call *Homo habilis* ('handy' or 'tool-making man'). Their simple tools, little more than pebbles with parts of the edges chipped away, are the oldest manufactured objects to have survived.

The basic needs

We know very little about how these tools came to be made, or what they were used for. Some have long, flat blades and look like scrapers; others with shorter, more pointed, cutting edges, seem to be choppers. Indeed most of the earliest tools were probably general-purpose implements. One can imagine many environmental challenges that could have stimulated tool-production and tool-improvement. Among the most pressing would have been obtaining and preparing food. Here stone implements would be invaluable for slaughtering animals and butchering their meat, as well as for cutting up plant matter or digging in the soil. Such tools would also have helped in cutting or chopping branches for creating shelters, and for scraping animal skins to produce basic clothing.

To begin with, there was probably little idea of making specialized tools to fulfil these different needs. The shape and size of the tool would have more to do with the proportions of the original stone, and with how it reacted to working with another stone. So one useful way of classifying early stone tools is not by their use but by the way they were made. At the simplest level, there were two basic types, core tools and flake tools.

Imagine one of our ancestors with a likely-looking piece of flint in one hand and a tough hammer-stone in the other. By hitting the flint with the hammer-stone, thin slices (sometimes quite large ones) will be removed from the flint. One can think of the large piece of flint as the core. If enough slices are removed in the right places, the core will be shaped into a tool of usable shape with a sharp edge. This will be a core tool.

But some of the slices will themselves have sharp edges and may also make good knives or scrapers.These can be salvaged and put to use as flake tools.

On the whole, the core tools tended to be heavier than the flakes: cores made good choppers and axes, flakes better knives or scrapers. And as the flint-worker's skill increased, this element of specialization would have increased too. But to begin with, the number of basic shapes was quite restricted.

Pebble-tool cultures

Archaeologists distinguish different groups of sites around the world according to the types of tools they produced. The earliest are the 'pebble tools' found in

For much of the neolithic, the variety of tools was dominated by the hand-axe. This all-purpose implement had two sharp edges. It was grasped in the hand, and could be used for cutting, grinding, scraping and pounding.

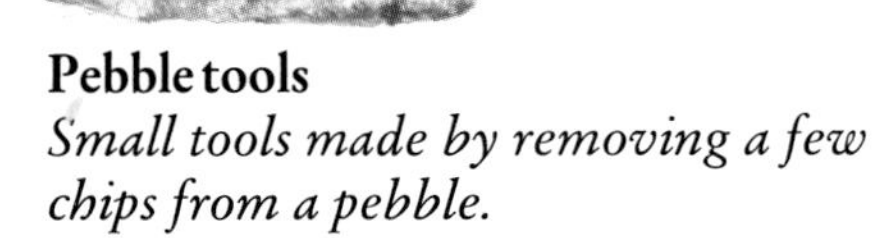

Pebble tools
Small tools made by removing a few chips from a pebble.

Blade tools
Still more minutely worked flakes of flint, for cutting or boring.

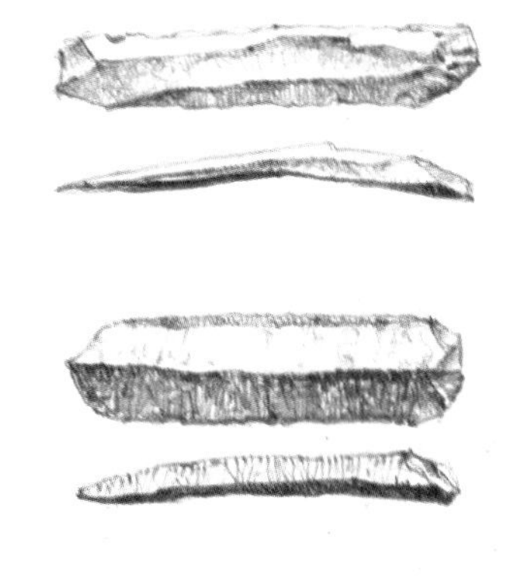

Core tools
More finely worked general-purpose tools made by removing many small flakes from a core of flint.

Flake tools
Flakes removed from a flint core, that have been worked to produce smaller cutters or scrapers.

East Africa, the most well-known of which come from the Olduvai Gorge in Tanzania, from a bed of rock that has been dated as 1.75 million years old. These are pebbles that have had flakes chipped from them, usually in two directions, to make a sharp edge. They could have been used for chopping, cutting and scraping.

Even this long ago, the inhabitants of East Africa displayed remarkable skill in their stone-working techniques. Most significantly, there is evidence from this site that tools were being made very deliberately and with forward planning. Archaeologists working on the site have identified stones that were probably anvils or hammer-stones, used repeatedly in the tool-making process. There is also evidence that the same methods of pebble-tool production existed in the region for hundreds of thousands of years. Indeed, they probably go back to the time of the first upright-walking hominids, some 4 million years ago.

The hand-axe makers

The makers of pebble tools gradually learned to work their materials more and more skilfully, developing the shape and the cutting edge until they came up with the two-sided tool that archaeologists know as the biface. This was usually oval or pear-shaped, and led to the production of the hand-axe, the form that replaced the pebble tool in Africa and then spread, along with the hominids who made it, into Europe and western Asia. Hand-axes seem to have been first made some 1.4 to 1.2 million years ago in Africa.

The typical hand-axe is almond-shaped, oval or triangular, and relatively thin in profile. Whatever the shape, one end comes to a sharp cutting point while the other end is broad to fit the hand of the user. The hand-axe was probably a multi-purpose implement, ideal for cutting, chopping and scraping. Thinner versions, with sharper points, may have been the ancestors of awls,which were used for piercing hides. Hand-axes also doubled as weapons, such as the tomahawk.

Most hand-axes were finely worked – increasingly so as time went on – the smaller flakes being chipped off with a piece of bone or a length of hardwood to give quite shallow scars on the surface of the tool. Hand-axes show one other key development: wherever the material was available, particularly in Europe, they were made of flint. This material was easy to work into a sharp, durable edge, and it became the favoured material of the toolmaker for hundreds of thousands of years. Indeed, the flint hand-axe seems to have been one of the most successful and versatile tools ever invented. However, the hand-axe did not spread to the Far East. Here, hominids living at sites such as Choukoutien (home of the so-called Peking Man) made do with quite basic choppers, similar in type to African pebble-tools, but rather more finely worked.

Flake tools

Toolmakers realized quite early on that small stone flakes could make useful tools. Once they had developed fine flaking techiques using bone and wooden hammers, they could make very fine implements out of these fragments of stone. Such tools were made extensively in Europe, where flake-tool traditions often existed near, or together with hand-axe-making cultures.

The simplest way to make flake tools was simply to strike a core of flint and utilize any likely-looking pieces that fell off. But people learned that hitting any old flint core would produce unpredictable results. So they shaped the core first, often working it into a shape like an upside-down tortoise shell. This would then be hit at one end in such a way that an oval flake would come free from the gently curving side. It would already have thin, sharp sides that would make serviceable knife edges, and a rounded shape suitable for scraping a hide. Sometimes narrow strips of flint would be split from a core in this way, flakes that would be more useful for cutting than scraping.

This method of making flake tools is known as the Levalloisian technique, after a site in the suburbs of Paris. The Levalloisian culture flourished in France and England around 230,000 years ago. Many tools made using the Levalloisian technique show little or no evidence of being worked after being split from the core – once the core had been prepared, it was possible to make almost instant tools.

Specialized tools

For the sake of convenience, archaeologists often speak about cultures producing either hand-axes or flake tools, as if the people created only one basic type. The reality was different. Hand-axe makers would make use of the stone flakes that

Tools and weapons, uses and materials
The toolmaker of the neolithic usually had to make do with locally available materials. Where there was no flint, it was either necessary to resort to less suitable stone, or to use bone or antler. All these materials could be fashioned into a variety of shapes, and sometimes it is not easy for the archaeologist to guess whether an item was a tool or a weapon.

1 *Bone – a universally available material.*

2 *Antler – a material that could be removed from the quarry of the hunt and carved with relative ease.*

3 *Stone – many types were used, but flint gave the sharpest, longest-lasting edge.*

4 *Serrated-edged flint knife.*

5 *Stone-headed axe with the head inserted in an antler socket to keep it secure.*

6 *Bone weapon.*

7 *Bone harpoon, attached to the shaft with twine.*

8 *Harpoon with bone prongs that open to hold the weapon in the wound.*

9 *Finely carved flint knife.*

4
5
6
7
8
9

were produced as they worked; Levalloisian people would also use core tools, making use of whatever implement best suited the job they needed to do.

Later, as the Palaeolithic period advanced, the making of specialized tools became much more deliberate. The most notable advance came in the Upper Palaeolithic period during the final glaciation, some 35,000 years ago. The people responsible for these developments probably originated in eastern Europe and south-west Asia, and spread west across Europe, keeping south of the great sheets of ice that stretched across the continent. They were nomads, and had to cope with inclement conditions and move on quickly when food was scarce.

It is perhaps in this period that, for the first time, it makes most sense to talk about a 'toolkit' – the people of the last ice age produced a great variety of tools and weapons, and mastered the art of working bone as well as stone. Their implements ranged from utilitarian knives to chisel-edged burins that their artists used for engraving on stone, bone, antler, or wood.

Archaeologists identify numerous different cultures in Europe at this time, all with different abilities and styles of toolmaking. The skills exhibited by the toolmakers of these cultures included the following: the ability to make efficient blade tools from sharp-edged flakes of flint; the ability to work bone by cutting, splitting and rubbing down; the ability to make narrow pins, needles and awls; the skills of hafting, allowing an object such as a blade or an arrowhead to be joined to a wooden shaft; the techniques of polishing stone to a shiny surface; the use of wedges for splitting wood and bone; and the art of 'pressure-flaking' to produce very finely worked tools and weapons out of flint.

These techniques were put to use in the manufacture of many different types of implement: knives and scrapers; burins and awls; axes and choppers; arrowheads,

Olduvai Gorge

Palaeontologist Mary Leakey searches for evidence of early hominids.

Millions of years ago the great plain of Serengeti in northern Tanzania contained a large inland lake. This lake was the source of water for many animals, including early hominids. Gradually, as a result of gravel and silt deposits and volcanic activity, the lake began to be replaced by strata of rock, which eventually built up to a thickness of hundreds of feet. Finally, a river cut its way through the rock, creating the valley we now know as Olduvai Gorge. Trapped in the rock at various levels are the fossilized remains of the creatures that once lived in the region.

At first it was thought that there were no human remains among this material. But this view was overturned as a result of the work of Louis and Mary Leakey between the 1930s and the 1950s. The Leakeys discovered a number of fossils of early hominids. These included Australopithecenes – ape-like creatures who probably share a common ancestor with humans, and a specimen that the Leakeys dubbed Homo habilis *– toolmaking or 'handy' man. This creature is thought by some to be a direct ancestor, by others to be a closer relative to the Australopithecines. Other important remains included* Homo erectus *specimens, and a skull discovered by the Leakey's son Richard. This hominid, known to science as ER1470, is some 2 million years old, but displays strong similarities to the human species. Layers of volcanic rock separate the layers bearing the hominids, and because such rock can be dated using the potassium-argon technique, the fossils too can be dated approximately.*

Tools of many different types were found in the rocks of Olduvai Gorge with these hominid specimens. The most basic are pebble tools, simple choppers that come from the same rock strata as the Australopithecines. These ancient tools have been called Oldovan, after the site. Higher up, contemporary with the Homo erectus *fossils, are more sophisticated tools – biface implements of various types with either straight or curved cutting edges. Palaeontologists call these tools Acheulean.*

These objects have been discovered in various contexts. Sometimes, when they occupy a very shallow layer, it can be deduced that they represent a former living area. Sometimes they are grouped around a large mammal skeleton and probably represent the site of a kill and butchering. But more often they are spread through a deeper layer, or are among debris in an old stream channel, and it is less easy to draw conclusions about the sort of life they represent.

spearheads and harpoons. Many of these opened up new opportunities. The burin, for example, was useful not only for artistic activity. It could itself be used in tool production, enabling more sophisticated flaking to produce items such as backed blades. These were knives with one edge left as a blunt spine, the other sharpened, like many knives today. The needle and awl would enable better clothes to be made; a greater number of more efficient weapons would make it more likely that the food supply would be maintained; better knives and axes would permit easier construction of homes to shelter people from the cold weather; and so on.

The inheritance of the toolmakers

Many of these early tools bear a close resemblance to metal hand-tools used until very recently. But what is more remarkable still, perhaps, is what these early tools tell us about the people who made them. And what they tell us is that these were indeed humans. They could not only look at a problem and find a way to solve it – they could also conceptualize and plan for the future, creating items such as hammers and hammer-stones which would be used in the production of many tools; and create tools for specific needs that would arise in the future – from butchering carcasses to engraving.

The stone tools also show us our first glimpse of mankind developing a technology – in the way that we will see many other technologies developed during the course of this book. This is a development that took place over hundreds of thousands of years, and saw solutions being developed independently at different points on the globe. It also involved the quarrying of raw materials – flint and obsidian – that represented one of the first entrepreneurial activities. These are phenomena that still typify the process of invention around the world today, and that show how basic such a process is to the story of human development.

Stone-working techniques

There were several ways available to early people of working stone to the required size and shape, ranging from simply banging two stones together, to the precise pressure-flaking techniques employed on later stone tools. Several methods might be used on the same tool. Sometimes, as with Levalloisian implements, a flake that was usable as a tool could be removed from the core in one operation.

Percussion
A stone pebble used as a hammer allows flakes to be removed from a tool.

Wooden hammer
Hardwood battens were sometimes used to remove flakes.

Pressure flaking
A bone burin with a sharp point was ideal for removing tiny flakes, producing an immaculate finish and a razor-sharp edge.

Levallois technique
When the core had been shaped with a number of precisely positioned strokes (2–5), a single blow with the hammer stone (6) could remove a sharp-edged, ready-made tool. The arrows indicate the direction of the flintworker's strokes.

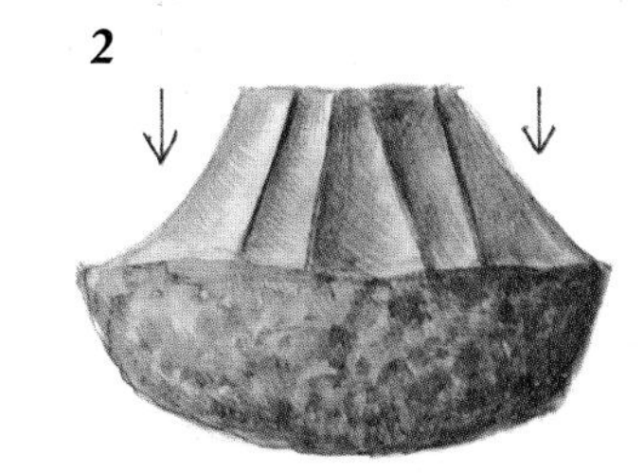

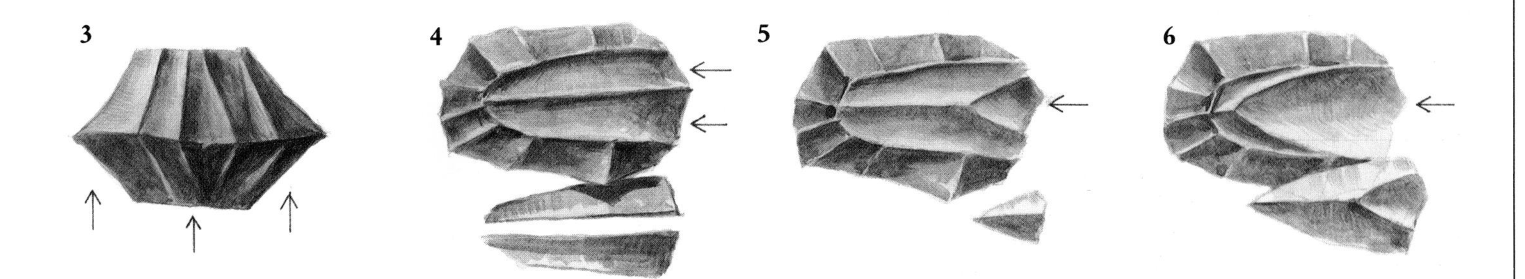

MAKING FIRE

The spark that kindled a new power and established the link between hearth and home

One of the most fundamental human skills is the making and the use of fire. As the hominids spread around the world, moving into areas where the climate was cold, fire became important for heating. It was also obviously useful for cooking (and this use would have extended the range of foods available to the hominids), even though a lack of evidence for fire at many early human sites suggests that these people ate their meat raw. Another use of fire was as a source of light. And there was one further use, less obvious today, but perhaps most important of all when it was first discovered. Fire afforded protection from the wild animals that attacked early humans; a continuously burning camp fire would keep predators at bay. By extension, fire could be used in hunting, to help flush out game.

The first encounters of people with fire would have been when it occurs naturally the result of trees struck by lightning, from surface fires at oil deposits, or even from volcanic activity. From such encounters people learnt the properties of fire: its heat and light, and the ability of dry materials like wood to catch fire.

From here it would be a short step to taking fire home, by making a torch with some wood and twigs and carrying fire from the natural site to the cave or camp. Here, once started, the fire could be kept going indefinitely, a constant source of heat, light, and protection: the symbolic link of hearth and home was established.

Probably the first creatures to use fire in these ways were not true humans at all but members of the species *Homo erectus* widely known as Peking Man. At the *Homo erectus* site at Choukoutien, China, archaeologists have discovered much evidence of the use of fire. Hearths, ashes, charcoal, and charred bones, all dating back to the Middle and Lower Pleistocene periods, 300,000 to 400,000 years ago, have been found at Choukoutien. In this period a large cave here was occupied by *Homo erectus* hunter-gatherers. There were ample remains of animals in the cave, mostly deer bones, and it is probable (although there is no way of knowing for sure) that these represented part of the diet of the occupants of the cave, and that the people cooked their meat before eating it.

It is likely that these Homo erectus people needed fire before they could occupy their cave at all. The Choukoutien cave mouth faced north-east, and, like most caves, it would have been a dark, cold, inhospitable place without fire.

What evidence is there for the use of fire in the rest of the world? In California, large Pleistocene 'fire lenses' have been found. Surprisingly, evidence in Africa, which we have learned to look on as the cradle of humankind, is quite sparse. Only a few of the many African sites from this period show signs of fire notably the Cave of Hearths at Makapansgat, South Africa, and the Montagu Cave in Cape Province. There is a similar paucity of evidence in Europe, though this may be because the damp climate has been less kind to charred organic remains. But Europe does show one important development. At Terra Amata, Nice, in southern France, remains have been found of a shelter made of branches, about twenty-six feet (eight metres) long, built on a beach, with a hearth inside. Among the debris of occupation were the bones of animals including wolf and giant ox.

Homo erectus

And it was not just a question of where you lived. With fire as a means of protection, small groups of men could range far and wide in search of food or new places to settle. Previously they had to travel in large bands to defend themselves. So fire had a double effect: it enabled greater mobility, while at the same time it reinforced the permanence of hearth and home. It is not surprising that fire became a vital part of the ritual and mythology of many early peoples.

Fire-making techniques

All the developments described on this page could have happened with fire that was found and transported. But it may not have been long before someone saw sparks coming from two branches being rubbed together in the wind, and had the idea of trying to start a fire by rubbing two sticks together. We know from studies of recent peoples living in traditional lifestyles that quite sophisticated devices have been produced to do this. One of the first was a round stick rotated quickly between the palms while being pressed down on a flat, wooden hearth. Later, bow drills were used to rotate the fire stick faster, and thus more quickly make the fire. And at some other time it was discovered that a spark could be created by striking a piece of iron pyrites with a flint.

Find a good way of rubbing two pieces of wood together and you can make fire. An early method was to turn a stick in the hands. Later fire-makers used the bow drill to turn the stick more quickly and press it hard against the wooden hearth.

RITUAL AND CEREMONY

Once the fight for survival had been achieved, early humans began to formalize rites of passage such as birth and death

The idea of identifying and marking the key events of our lives goes very deep. It is something we see at all levels of society, and in all types of society, from initiation rites in traditional cultures to opening ceremonies for modern schools or hospitals. In the world of our ancestors, similar milestones would be celebrated – the birth of a child, its coming of age, the linking of partners in the ancient equivalent of marriage, the last rites of death. Events connected with day-to-day survival – success in the hunt, or the coming of spring – might also have been celebrated.

For the archaeologist, the type of ceremonial for which there is most clear evidence is the sort connected with the disposal of the dead. No one knows when this was first turned into a ceremony. Perhaps the earliest hominids simply took their dead away quietly to some place far from the main settlement site and left the bodies in the open: scavenging animals and the elements would soon do the rest. But at some point, perhaps around 80,000 years ago, people began to bury their dead with obvious ceremony.

Here is one example. At Teshik-Tash in Uzbekistan, a boy some 8 or 9 years old was buried. His body was placed in a shallow grave and his head surrounded by five or six pairs of horns from a Siberian ibex, placed upside-down in the soil to form a crown. The grave is in a cave, high in the mountains, in an area where these goats are still common. The animals may have been an important source of food and skins for the people of the region, hence the use of horns. But what is clear is the care with which the child was buried. Death was being treated ceremoniously.

This is a pattern which is repeated in many burials that have been excavated from the period of the last glaciation and just afterwards. This was the period when our species, *Homo sapiens*, was established in Europe, both as *Homo sapiens neanderthalis* (Neanderthal man) and our own subspecies, *Homo sapiens sapiens*. So, ironically, the Neanderthals, crude and thick-skinned in the popular imagination, may have been the first hominids with the refinement to create ceremonies and bury their dead with dignity.

Such sophistication in ceremony is undoubtedly related to sophistication in other areas of life. At this time people were facing great challenges. There was the need to survive in a climate that was cold and inclement – to find food and to keep warm. These challenges probably led to a desire to control the environment and to understand it. In such a context, death becomes one of life's mysteries. Why do people die and what happens when they do? It is not surprising that burials started to be treated seriously, and that items relating to the environment were included in the ceremony.

These early burials themselves hold many enigmas for us now. There are not enough sites to enable one to generalize about how the dead were treated at this time. Clearly, not everyone was buried, and we do not know why some people were buried and others were not. What is more, the bodies are treated in different ways, being positioned differently in their graves. Also, grave goods such as stone tools are sometimes included, although this is unusual in the oldest burials. The inclusion of grave goods in later burials is usually taken as indicating belief in some sort of afterlife: the tools, vessels and so on are meant to be of use by the deceased in the next world. But there is no evidence of such beliefs in the period of the Neanderthals. Finally, we cannot know what actually took place at these early ceremonies. There may have been ritual dances and music played on skin drums, wooden percussion instruments or bone pipes. The origins of music and dance might be connected with ceremonies like these, but we do not know for sure.

And yet such remains do give us great insights into the lives of the people who left them. For example, in a grave at Shanidar, Iraq, is buried a man with a deformed arm and shoulder, the result of an arthritic condition he had probably had since birth. Yet this man lived to an age of about 40 – clearly he was cared for by his companions, even though he could not have provided his own food.

It is the importance of things other than the purely utilitarian, the idea of life being more than mere survival, that the evidence of early rituals brings home to us today. As such, it is a key element in our make-up, an indication that great ideas come from the challenge of understanding the world as well as from the basic need to keep on living.

Burial site
The presence of grave goods and the special positioning of the corpse testify to the care and attention lavished on even the earliest burials.

Dance, music, bodily adornment and trance-like states were probably all part of early rituals, although there is no way of being certain about the content of such ceremonies. But we can be sure that ceremonial sites were accorded great importance. Massive ritual centres like Stonehenge in southern England must have taken enormous effort to build.

ART

Although techniques have changed, art still serves to depict ideas and express emotions, just as it did for our ancestors

When we think of the most effective ways we have evolved to express the human condition and to comment on the world around us, we think straight away of the arts. Indeed, so close is the affinity between art and humanity that some authorities use art as the key indicator of human evolution: when we become artists we become truly human for the first time. (However, it should be noted that some people regard the efforts of chimpanzees and elephants as art.) Perhaps the identity of art and humanity is to do with the fact that to be an artist one must look at the world in a particular way. The artist must stand back, contemplate a situation, make a set of decisions about how the subject is to be represented, and record the results of these thoughts in a way which will communicate them to other people on a level somehow above the mundane. This balance between thought and communication makes up a process which, perhaps more than any other, involves the 'great idea'.

As one would expect from such a basic human activity, art seems to have arrived on the scene at just the time when modern humankind was evolving – during the Upper Palaeolithic period (35,000 to 8000 BC). By this time humans had spread over large areas of the globe – from Africa to Europe and Asia, and thence into the Americas and Australia.

In Africa, where, according to most anthropologists, the story of humankind began, a certain amount of Palaeolithic art has survived from the early rock shelters. Examples include painted stone fragments dating back some 40,000 years. America also has early art, but this has proved very difficult to date. It includes rock inscriptions, and the finely decorated pottery of the Indians of the southwest. Australia has some impressive examples, certain of which have proved easier to date because decorated rock surfaces have been partially buried beneath debris from an ancient period. For example, engravings in the Early Man Shelter in Queensland disappeared beneath layers some 13,000 years old, proving that they must have been made before that date. Other examples in Arnhem Land, northern Australia, were buried under even older layers – up to 30,000 years old in some cases.

Asia also has examples of Palaeolithic art, but it is Europe that has yielded the most, and the most beautiful examples of the beginning of our artistic heritage. This art takes several different forms. There is a variety of portable pieces (engravings on stone and bone, carvings in stone, bone and ivory), engravings on the walls of caves, clay models and, most famous of all, paintings on cave walls. In terms of subjects there are numerous apparently abstract patterns (dots, lines, dashes and other marks), portrayals of humans and, above all, depictions of animals (horse, bison, deer, aurochs, ibex, mammoths; even big cats, bears, fish and birds). Much, though by no means all, of this material is of superb quality, still speaking to us eloquently today, even if we do not know enough about the circumstances of its creation to understand it fully.

'Dreamtime owl' from northern Australia

This quality tells us that there is something special about this art. There is nothing mundane about it and the position of many of the paintings deep inside caves where access is difficult and light hard to come by increases their impact and their mystery. How were the images produced, and what can we learn from them about the significance of art at the beginning of the human era?

Producing the art

One of the most common Palaeolithic art forms is the engraved stone. Thousands of examples have survived, bearing a variety of designs, from quite simple incisions to detailed animal portraits. Some of the incisions are very shallow and are visible only in strong sidelighting – a fact that caused many examples to be overlooked by early archaeologists.

The shallowness of some of these early traces may be due to the inadequacy of the stone scraping tools that the artists used to incise the hard stone. But there is evidence that the engravers possessed quite sophisticated tool kits for their art. Microscopic examination of some pieces suggests that burins with several different types of points were used to achieve varying effects of line.

Bone and antler were also engraved in this period – and sometimes also carved. These materials proved easier to work than stone. A range of tools, from small stone blades to awls and burins, could be used to work bone, and some beautiful carvings of subjects such as bisons and mammoths have survived, some decorated with ochre pigment.

There are some magnificient antler carvings, such as the famous spear thrower decorated with a young ibex from Mas d'Azil in the Pyrenees. Such figures, which come from the Magdalenian period of the Upper Palaeolithic, show the remarkable development that human art reached between 15,000 and 10,000 years ago.

So too do the works on cave walls, particularly those in France and Spain. Lascaux and Altamira are perhaps the most famous of these, but there are many other decorated caves – some with work as good as the better-known ones. Again, the most frequently found technique is engraving, often shallow and difficult to see, especially when the rock is hard and the light poor. A strong, sizeable tool would have been needed, with perhaps a smaller flint burin for the finer work. Where the rock was so hard that conventional engraving was impossible, the artist sometimes scraped away areas of the surface to reveal stone of a different colour, a technique more like painting than engraving.

Ancient art today
Much of the art of the Ice Age is engraved in shallow lines on the walls of caves. Sometimes one line is incised above another in a bewildering criss-cross. Amidst the complex designs we can sometimes discern the figure of a horse or mammoth. We know these drawings largely from the copies of archaeologists, and their clarity depends largely on the way the copyist has selected which lines to reproduce. This is another reason why we should always be cautious in our interpretations of Palaeolithic art.

Alternatively, artists could abandon the rock surface altogether and model in clay. The best-known example is the pair of bison in the cave at Tuc d'Audoubert. About one-sixth life size, they lie in a low-ceilinged chamber, to which clay had to be brought from another part of the cave. Sausages of clay still on the floor bear the sculptor's finger- and palm-prints – no doubt the artist was testing the malleability of the clay while deciding whether to use it for another part of the sculpture. The bisons themselves are in bas-relief, although their posture and their deep modelling make them seem almost fully three-dimensional.

The final method of decoration, painting, was very widespread. The palette was quite restricted. Most artists got their pigments from natural ochres and were content with a range of colours encompassing red, yellow, brown, black and, occasionally, white. Locally available water was used to bind the pigment, which was then applied with the fingers or with animal-hair brushes or pads. With such simple means, in deep caves that must have been well out of the way of regular human traffic, the painters of the Upper Palaeolithic created some of the most remarkable images in the whole of human art.

Animal hunt cave painting

One of the most interesting things about these paintings is how simple they are. The number of lines is relatively small; the palette quite limited. Modern artists trying to imitate the paintings of Lascaux found that they could produce something similar very quickly. It is not difficult to produce such figures in well under an hour, including preparation – albeit without the distinctive touch of the Palaeolithic master.

The significance of the art

This improvised, almost sketchy quality has led many people to see cave paintings as resembling modern art. But we should not be drawn too far into the trap of thinking of Palaeolithic art in the same way in which we think of the products of the modern artist's studio. Some of the early discoverers of cave painting in the late nineteenth century tended to think of it rather in this way: as the decorative pastime of people with time to spare – art for art's sake, in fact.

But clearly, life in the Palaeolithic was very different from the life of the educated middle classes of the late nineteenth century. Consequently, people began to look for a more plausible set of meanings for cave art, one more relevant to life in the stone age. Many commentators found this in the idea of 'sympathetic magic'. According to this theory, the effect resembles its cause in some way. So, a picture of animals might be drawn to bring about success in the hunt, and give heart to men anxious about the availability of food.

The paintings themselves might even have been part of some pre-hunt ceremony: they may have been created during the ceremony, or missiles might have been thrown at the painted creatures in a pre-enactment of the hunt. Many paintings show signs that they have been treated in this way. Some paintings also have lines marked on them which have been interpreted as arrows.

These ponderings provide an attractive theory because they seem to get to the heart of the early humans' struggle for survival, while also explaining the subject matter of the art. Another version of the 'sympathetic magic' theory sees the paintings as fertility images. The profusion of animals is drawn in the hope that the real creatures will multiply to provide a rich source of food for the people. However, both these versions of the same theory rely on careful selection of the available evidence.

An important step in the interpretation of cave art came when scholars began to look at the whole group of paintings in a cave, rather than individual images. This led to some interesting perceptions. For example, rarer animals, such as bears, big cats and rhinos, are often in the more remote inner parts of a cave, while creatures such as deer, ibex and mammoth are near the entrance, and horses and bison are found on large central paintings. Similar generalizations could be made about the arrangement of male and female subjects. Hence another theory based on analysis of whole caves suggested that the images were a mirror of the peoples' social organization, with different animals representing the mythic ancestors of various family groups.

Such theories have alerted us to the fact that there is no simple explanation of cave art. It could conform to one or more of the numerous theories that have been put forward, or even to some as yet unthought-of theory. But the very fact that it is so open to interpretation, that, often hidden away in inaccessible places, it has that special quality that makes us want to go on looking at it, confirms that it is true art, and the beginning of a typically human mode of expression – the product of typically human ideas.

Venus figurines

Although much of the art produced in the gloomy setting of the Palaeolithic caves consists of animal representations or abstract patterns, the artists of the Ice Age and their successors produced some portrayals of the human figure. Among the most famous are the so-called 'Venus figurines'. The large buttocks and breasts on some of these carvings have led prehistorians to think of them as fertility figures. But a study of a large number of these sculptures has shown that fewer than 20 per cent have the large breasts and fleshy body of the 'typical Venus', while rather more have young, flat-stomached bodies, or seem to be portrayals of older women.

The figures remain rather mysterious because many were discovered before archaeologists realized the importance of recording the exact place (and level) where such items were found. Some of the eastern-European figures, however, have been found in special pits made in what were the floors of huts. This has led archaeologists to suggest that they represent revered ancestors. Whatever their exact meaning, the care spent creating these figures and placing them in their pits suggests that these societies may have been ones in which women were held in particularly high esteem.

Types of ancient art

Using sticks to mix their pigments and often applying their paints with their bare hands, the artists of the Ice Age produced some of the greatest images. A single cave sometimes contains a whole range of subjects superbly rendered. Some of the variety of ancient art is indicated by the small pictures across the top of these two pages. In Europe alone, many different human and animal forms appeared. From left to right are shown:

1. *Wall painting of charging bison, Altamira, Spain, c.20,000 BC*
2. *'Venus' of Willendorf, carved in limestone, Germany, c.30,000 BC*
3. *Female figure, carved in ivory, central Europe, c.20,000 BC*
4. *Head of girl, carved in ivory, Brassempouy, France, c.20,000 BC*
5. *Carving of obese female figure, Germany, c.20,000 BC*
6. *Staff or spear thrower, carved in antler, southern France, c.20,000 BC*
7. *Stone statue found at St Sernin, France, c.3000 BC*

IMPROVING MAN'S CHANCES OF SURVIVAL

Survival: the most basic motive for human behaviour. And some of the most important human inventions are responses to this fundamental need. Food, health and shelter are perhaps the three key areas – keep these needs satisfied, and life, both of the individual and the species, can go on. It is not surprising, then, that people have been obsessed with them since the very beginning. It has been said that no arts have changed the face of the earth more than agriculture and architecture; and it might be added that medicine has the greatest potential to improve life.

All of these ideas are rooted in our most remote past. Their origins are obscure, because from the very beginning people had to promote their own survival. From the word go, people would notice which plants made them ill, which were good to eat, or even made them feel healthier. As soon as people began to live anywhere with an inclement climate, and where there were no caves to provide protection, artificial shelters must have been necessary.

From such basic information, the skills of horticulture, medicine and building were gradually amassed over thousands of years. Facts and ideas about food plants, medicinal herbs and building materials were tried, tested, and passed on from generation to generation. People were piecing together evidence about their environment, and then beginning to use it to their advantage.

As the early nomadic peoples wandered the earth, they must have amassed a formidable body of information about the environment. Such knowledge is clear from the sophistication still prevalent in traditional societies today, in fields such as herbal medicine. Their knowledge took another turn when some of the nomads began to settle down, to grow their own crops and breed their own animals, and to become so efficient at making food that they could produce a surplus.

This led to specialization, to deeper and more objective research, to greater co-operation and information exchange, and to civilizations where written records of scientific discoveries could be kept. In such an atmosphere, disciplines like agriculture, architecture and medicine took on a life of their own. And so buildings would be produced that provided more than mere shelter (they became status symbols, or works of inspirational art) and yet more efficient methods in agriculture and medicine would evolve.

This process has continued for thousands of years. It has taken agriculture, architecture and medicine far from their early origins. A Mesopotamian mud-brick house seems a long way from a modern office block, a Chinese herb boiled in water and drunk a far cry from one of the synthesized drugs of today. And yet as modern architects take inspiration from traditional building types and today's chemists analyse traditional medicines and find their active constituents, we can see what the science of yesterday and today have in common: a curiosity about the environment, and a commitment to using it to improve our life on earth – and, it is to be hoped, to benefit the earth on which we live.

EARLY FARMING

Instead of travelling to find food, humans began to settle down and produce it themselves, thus beginning to control the environment rather than be controlled by it

Until the end of the ice age most, if not all people probably lived by hunting and gathering. When green leaves and fruit were accessible they would pick these and eat them; in times of scarcity, they would hunt for meat, trapping or spearing whatever animals they could find, from small mammals of all sorts, to great woolly mammoths. Naturally, the animals hunted depended on what was available. In northern Europe there were reindeer-hunting cultures, while in northern Africa some groups survived by hunting the barbary sheep.

During this late-glacial period (from about 23,000 to 14,000 years ago) conditions were both colder and drier than later – drier because the large amount of water locked up in the ice-caps reduced the extent of the seas and thus the amount of evaporation from them.

Changes in the environment

But with the end of glaciation, about 13,000 years ago, things began to change. The temperature increased and trees began slowly to spread into the new warmer, damper areas – birches and conifers were followed by slower-growing species such as oaks. And with the trees came animal species such as deer, elk and wild pig. Numerous lakes were left where glaciers had been, and these were colonized by waterfowl and fish. It was a welcoming environment for people too, who began to move into new areas, to prosper and to increase in numbers.

It was at about this time that people began to realize that they could provide food for themselves in a different way. They could harvest the seeds from plants and cultivate them in a place favourable for their growth; and they could begin to domesticate animals to provide a reliable supply of meat. For the first time, humans were becoming farmers.

No one knows exactly where or when this began, or what was the decisive reason why people started to organize their lives in this way. Our best evidence for early agriculture comes from the eastern Mediterranean (although it must have started independently in many different places). Here was an area that was already extensively settled. It was also a region of abundant large-seeded grasses, plants that could form an invaluable food source, that could be cultivated and harvested with ease, and that would evolve quickly to produce strains even more suitable for cultivation.

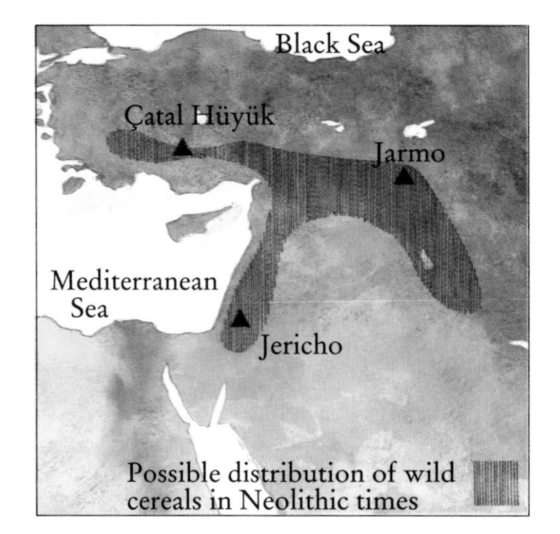

This map of the eastern Mediterranean shows the places where farming probably first took hold.

The coming of wheat

What the early farmers needed was a grain that produced abundant seeds. There were several varieties of wild wheat in the Middle East that did this, but they had disadvantages. To encourage seed distribution, the seedheads of wild wheats shatter when the seeds ripen. Ideally, the seedheads of cultivated varieties should not do this, since one wants to harvest all the seeds. However, hybridization created wheats that worked better for the farmer. Cutting with a sickle, when the head is less likely to shatter, and then sowing away from the wild population, encourages natural selection in favour of the non-shattering strains. And this is just what happened in the Natufian culture, which takes its name from Wadi-en-Natuf in Palestine. Here, reaping knives – straight, bone implements with a slot into which flint teeth were inserted – were used in some of the earliest-known examples of cereal cultivation.

As a result of developments like this, communities based entirely on agriculture had evolved in the eastern Mediterranean by about 8000 BC. Even the small-scale agriculture practised here could support larger populations than hunting and gathering. And as time went on, the growers became so efficient that they were able to produce more food than they needed themselves. These facts were to have profound consequences for the development of humankind.

Expanding communities

The growing agricultural communities of this period have left ample evidence of their existence. They built villages of mud-brick houses which were densely packed together next to their agricultural land. As time went on, new houses were built on the foundations of the original ones, and gradually a mound built up as successive generations rebuilt their homes on top of those of their ancestors. Such a settlement mound is known in Syria and Palestine as a 'tell', and in Anatolia as a 'hüyük'.

The most impressive of these remains shows that these early people were beginning to form societies so large, so capable of large-scale building projects such as stone walls, and so able to support specialist workers who did not need to grow their own food, that their communities could earn the name of town. And probably the first town, on the west side of the River Jordan overlooking the crossing to the north of the Dead Sea, was Jericho.

Types of wheat

The naturally occurring species of wheat that were found in the Middle East provided nutritious food but were difficult to harvest and use efficiently. The seedheads tended to shatter when they were ripe (a natural mechanism to help seed distribution) and they were difficult to thresh. Early farmers found that related forms would hybridize to produce some strains that had non-shattering seedheads, and which were also easier to thresh. So they selected and encouraged these forms, such as einkorn, emmer and macaroni wheat, which were commonly grown in the eastern Mediterranean by the first farmers.

The most important species

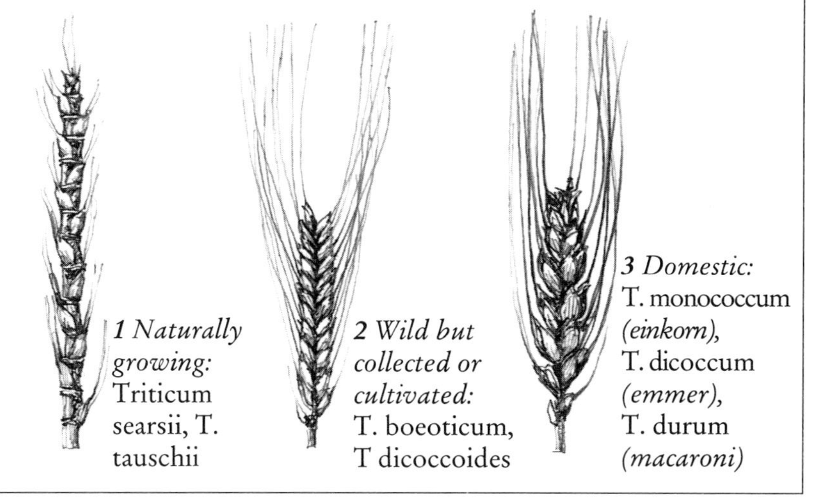

Over the centuries, Jericho would be followed by other towns around the eastern Mediterranean and into Anatolia, where the settlement of Çatal Hüyük dominated a large area. With agriculture producing a surplus of food, craft workers could specialize, producing a range of objects from tools to pots. And these objects could be traded. One precious material exploited in this way at Çatal Hüyük was the natural volcanic glass called obsidian. This was prized for its usefulness in the manufacture of stone-cutting tools, and tools from the Çatal Hüyük region turn up over a wide area of Anatolia.

So agriculture helped to encourage the growth of towns; towns fostered trade; and trade allowed the townspeople to dominate wider hinterlands around their settlements, and thus to control a still wider environment. The rise of civilization, the growth of empires, the expansion of human horizons – all of these are linked to the coming of agriculture, as are the domination of one people by another.

Typical early farming community

Fields of cultivation

What were the crops and animals that enabled these developments to take place? As we have seen, the different strains of cultivated wheat were vitally important. But there were other crops – the idea of growing several different plants to avoid disaster if one crop failed seems to have been a very early one. Thus barley was grown alongside emmer and einkorn wheats, and in some areas, such as Syria, lentils provided a nutritious crop.

The choice of livestock also depended on what was available locally. In the Natufian culture, for example, remains of gazelle and goat have been found. This evidence, together with the fact that many of the remains come from immature animals, suggests that these creatures were herded. At some point, there was a change from gazelle to sheep and goat, and later still, the animals typical of modern farming, sheep, cattle, and pig, were used much more widely.

The multi-layered sites of the tells and hüyüks sometimes show the process of change. One example is the site of Çayönü Tepesi on the southern side of the Taurus Mountains. Lower down, on the earlier levels, are remains of deer and wild cattle, pigs and sheep; the evidence for grains and legumes shows them to be wild species also. On the higher, more recent levels, the emphasis has changed, and there are pig, sheep and goat remains that look more like domesticated varieties, together with cultivated emmer wheat and pulses.

Improvements and developments

By the sixth millennium BC, technology had developed in the eastern Mediterranean to the extent that agriculture was widespread, and pottery was being produced, some of it of high quality. In addition, metalworking had begun in many areas, with the production of beaten-copper pieces, causing archaeologists to call the period 'Chalcolithic', after the Greek word for copper.

The agricultural developments now began to spread further afield. The people of the Balkans, the Caucasus area between the Black and Caspian Seas, the Iranian Plateau, Turkmenia, and Fars, north of the Persian Gulf, all started to explore the skills of agriculture at about this time. And agriculture began to spread east and west, further into Europe and Asia.

Some of these new farmers met problems. In the lowlands of Iran and Turkmenia, for example, the climate was too dry for abundant crop production. So irrigation began, to take water from the mountains to the arid lowlands beside them. Simple irrigation techniques may also have been used on the Konya plain in what is now Turkey.

And even where agriculture was already well established, such as along the twin rivers of Mesopotamia, the Tigris and Euphrates, techniques were developing. Here hoes of ground stone began to replace cruder flint tools as land that required more intensive preparation was pressed into service.

Why cultivate?

Everyone who has thought about the beginnings of agriculture has asked why it happened. After all, there were many parts of the world where people got along quite happily without agriculture until very recently. People would adapt to a locally available food that could be gathered easily, or move on until they found a place with such a food. In the plains of North America, for example, great herds of bison provided the Plains Indians with a seemingly endless supply of food, as well as hides for clothing. Such people did not need to farm, and kept their nomadic lifestyle.

There is even an argument that points out the disadvantages of agriculture. In particular, agriculture is risky: if you concentrate a population of plants, animals or humans in one place, you increase the likelihood of loss due to weather or dis-

The first farmers
With the advent of organized cultivation came a number of tools and techniques for harvesting, preparing and cooking the produce. Many of these were adapted from similar methods used when food was obtained by gathering.

The flint sickle
Flakes of flint could be sharpened and inserted into a wooden handle to form a sickle for cutting corn. Straight forms soon evolved into the familiar curved shape. Flint could also be carved to make a serrated edge.

1 2 3

Early crops
Among the first crops were barley, grown in China (1); emmer wheat, cultivated in the eastern Mediterranean and Egypt (2); and other types of wheat, grown in Europe (3).

Grinding
Stones called querns were used to grind the corn into flour, from which bread could then be baked. There was a large lower stone with a depression for the grain to sit in, and a smaller stone on top. The small stone could easily be moved around by hand. Corn was ground like this until the arrival of wind and water power – indeed, the technique is still used in some places today.

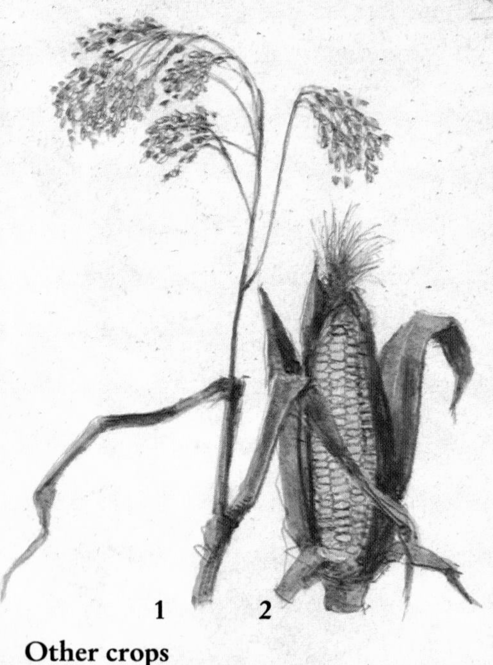

Other crops
Different crops flourished in some areas. Millet appeared in Europe in the Neolithic (1), and corn was a staple product in the Americas (2).

Winnowing
To separate the grain of the wheat from the outer chaff, early farmers made use of the fact that the chaff was much lighter than the grain. If the wheat was tossed in the air, the chaff could be blown away.

Baking
Bread – usually unleavened – has been a staple food for thousands of years. It was originally baked in clay ovens like this.

Storage
Pits lined with stones could provide cool storage in dry areas like the eastern Mediterranean.

ease. In other words, agriculture is a less reliable source of food than it seems. It brings further disadvantages in the form of drudgery (or even slavery), overcrowding and disease. Why did the people of the Middle East, and later those of Europe and the Far East and other areas, react differently to the problem of obtaining food?

According to one argument, the opportunity seemed too good to miss: the apparent reliability of agriculture made it attractive. Another view involves population pressure. The number of people was rising, especially in the widely settled Middle East, and the natural food supply did not provide enough to go round. This state of affairs might have prevailed in some isolated cultures, such as islands, where it is difficult to import food from the outside. If native animals had been hunted to the brink of extinction, there would be obvious sense in trying a way of managing the food supply more carefully, of growing crops systematically and domesticating animals. Another school of thought sees humans gradually 'settling into' their environment, coming to understand it better, and realizing that they will survive more easily by extending the natural food supply. And yet a fourth theory imagines people, instinctively drawn to areas with rich resources, overexploiting these resources to such an extent that farming became a necessity if they were to survive.

Power and the land

It may also be that the rise of agriculture is closely related to a change in the social and political power structures of early societies. Developments in craftsmanship and the exploitation of resources such as obsidian, and later metals, meant that control of such materials and the products made from them afforded power.

From such a situation, trade started to play an increasing part in the lives of many people. To trade you need to travel to get to new markets. But you also need bases from which to operate and places to store the goods that you are trading. Hence the rise of trade and the growth of settled communities and towns generally go hand in hand.

Large, concentrated communities need feeding, and if there was a perception that agriculture would provide a more reliable food supply, this might have been a cause of the rise of farming in the Middle East. Alternatively, local rulers and newly rich traders might have realised that farming offered potential for another sort of trade – trade in foodstuffs – giving them still more power. What is more, place and power go hand in hand for both trading empires and farming communities. The livelihood of both merchant and farmer is linked closely to specific places – market and warehouse for the merchant, fields for the farmer. Thus another key link between agriculture, trade and power can be established.

Settling down

All of these theories involve the idea that settling down and living in a confined area goes hand-in-hand with agriculture – there is a marked contrast between the wandering hunter and the sedentary farmer. Perhaps this settling down gives the key. People may have realized that a settled lifestyle gives safety in numbers, the sort of safety from which one can set out and dominate large areas of countryside, and large numbers of people. The attractions of trading and empire-building went hand-in-hand with agriculture, and, although agriculture itself has its drawbacks, the other opportunities it brings could have helped it to catch on.

Cultivation started independently in different places, and diverse factors may have been at work at different points around the globe. It used to be fashionable to talk about ideas spreading around the globe along trade routes. The notion that ideas start in one place and spread gradually in this way is known as diffusionism. Clearly, as agriculture was connected to the rise of trade the new science could easily have spread in this way. Trade itself created the pathways along which information (not to mention seeds of successful crops) could travel. No doubt, such a transfer of information did take place to a certain extent. In such a scenario, farming spread east and west through Asia and Europe together with the obsidian tools and pottery figurines that seem to have been traded at this time. But it is just as likely that discoveries about agriculture were being made independently in these places. As indeed they must have been in America, where crops such as corn were exploited without any contact with the peoples of the Old World.

In addition, cultivation may go back far further in time than our archaeological record shows. The kinds of animal and vegetable remains that are necessary to tell us about the origins of farming are not that common. It is probable that many earlier remains have perished for ever. Whatever its origins, agriculture has left an unique imprint on the world, and has been singularly influential on the story and development of humanity. Except in a few traditional cultures, agriculture provides the food that keeps us alive.

Water and irrigation

Agriculture began in parts of the globe where water was frequently in short supply. This often meant that quite elaborate measures had to be taken to irrigate the fields. Some of these projects, such as the digging of ditches or canals, could only be achieved with a large labour force and with much strategic planning. In other words, they needed a centralized government, and were part and parcel of the development of civilization. But such were the advantages to rulers of an efficient agricultural system and a food surplus, that projects like these were often undertaken.

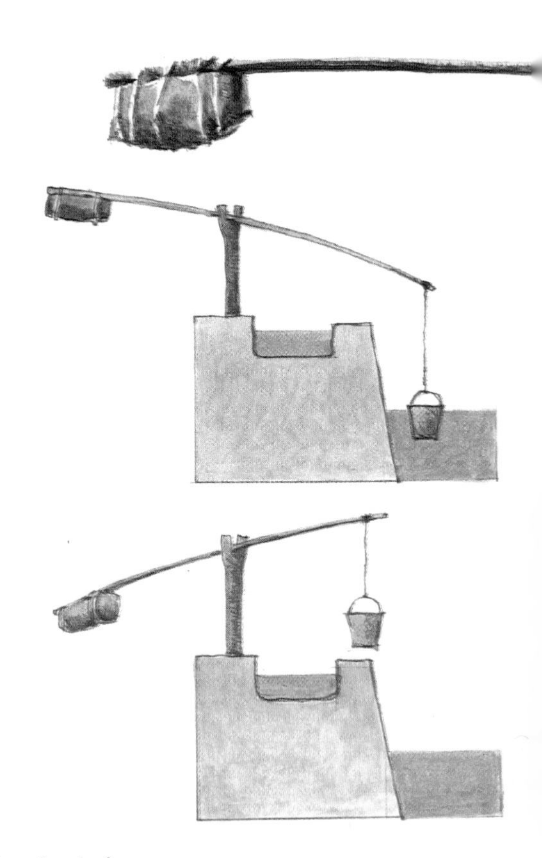

The shaduf

Raising water from one level to another was often a problem in early irrigation schemes. One solution was the shaduf, a bucket suspended on one end of a counterweighted pivoting pole. The Babylonians had shadufs at least as early as 2300 BC, and they are still in use today.

Raising the water

In the background a man uses a shaduf. With this device he could raise up to 600 gallons (2,730 litres) of water per day. The boys in the foreground work a treadmill to raise water with a system of paddles.

Rice, Oryza sativa, *another staple crop*

THE PLOUGH

An ancient invention refined by the Chinese around 600 BC, but which remained unchanged in the West for another 500 years

The first people who grew their own food did so on a very small scale. Some prehistorians describe this activity as horticulture (garden cultivation) rather than agriculture, and this gives an idea of what was involved. The tools used were primitive, and the plots compact areas. Often the land to be cultivated was an area of forest that the farmer had cleared by burning. The resulting ash would be scattered evenly over the plot and the soil beneath broken up with a simple digging stick. Then the farmer would sow his seeds, continuing to till the soil with a bent stick that acted as a basic hoe.

Eventually, someone noticed that dragging a stick along the ground was an effective way of breaking up the light soils of the Middle East. And a bigger, heavier stick would do the job still better, if you could get enough power to pull it along. Perhaps before long, an early farmer found a branched bough that could be pulled by two people, while a third walked behind, steering the end, and making sure the point was thrust well into the soil. The first plough had appeared.

No one knows exactly when this first happened. It was the kind of discovery that could have been made in many different places where people were tilling the soil. But there are carvings and seals from Uruk and Babylon in Ancient Mesopotamia dating from the third millennium BC that show such primitive ploughs in action. They had probably already been in use for centuries.

Beasts of burden

Cultivation with this type of scratch plough would still be quite a small-scale affair. And it depended on the right kind of soil. Scratch ploughs were of little use in the north, where soils were generally much wetter and heavier, nor in large parts of Asia. But one improvement was to come which would make ploughing an easier and more widespread activity.

At the same time as they were developing the use of crops, the early farmers were beginning to domesticate animals. The ox was the creature that first emerged as a suitable animal to do heavy work that was difficult or impossible for men. Harness oxen to your plough, and you would be able to work a larger area of ground more quickly and produce more food. You could realize the true potential of the science of agriculture in a way that was impossible before.

Again, it is from Mesopotamia and the Middle East generally that our evidence comes for this development. By the time of the Egyptian pharaohs of the Middle Kingdom (*c.*2130-1777 BC), tomb models of oxen drawing ploughs were quite common. The great civilizations of the Middle East – Sumeria, Babylonia, Egypt – were sure of a regular food supply.

Eastern developments

But there was one great civilization, often ignored by historians in the West, that contributed independently to the history of the plough. Stone ploughshares have been discovered in China that date to the fourth or perhaps even the fifth millennium BC. No doubt many wooden ploughs were also used in China in these early times, and certainly the tradition continued – Ancient Chinese writings of the fourteenth century BC mention ploughing.

But the development that set China apart from the rest of the world was the early discovery that iron could be used to make ploughshares. Iron had a good combination of weight and strength, and it could be formed easily into the right shape to cut and turn the soil. Iron was heavier than wood, but an iron share cut its way through the soil with less friction than a wooden one. The result was that an iron plough was both easier and quicker to use; it was also much easier to plough heavy soils.

This illustration is taken from a neolithic rock engraving of a scratch plough – an indication that the plough was present in very early cultures indeed. Although the engraving does not show the type of harness used with any clarity, it does indicate the share, with a handle projecting upwards at right-angles, which the operator would grasp to guide the plough in a straight line along the furrow.

(Right) *The earliest and most basic tools for cultivation were probably simple sticks. These could be used to make holes or shallow furrows to plant crops. Such digging sticks worked adequately, particularly in the light soils common in parts of Africa and the Middle East, although even here a stone axe might be needed to break up the soil. The plough, even a simple wooden one like that shown, was thus an important advance. It was quicker to use and could make deeper furrows in heavier soil. And when pulled by animals it was not such hard work for the human labourer.*

Iron was being used for ploughshares in China as early as the sixth century BC, about 500 years before iron ploughs appeared in the West. In some designs the whole share was made of iron, in others iron was laid over wood. Both designs would have been more effective than the wooden types being used in the West at this time.

The Chinese also developed casting techniques, enabling them to make a less brittle form of cast iron. By the third century BC, ploughs using this type of iron were quite sophisticated in design. They had well-sharpened points, and wings that flung the soil well to either side of the furrow. The result was a deeper furrow than before, and, once more, a plough that was ideal for use in heavy soils. And as time went by, Chinese shares got still bigger and better.

Perhaps just as crucial as the use of metal was the development of the mouldboard. This is the twisted plate above the ploughshare, which turns the soil over and makes it fall to one side of the furrow. We do not know exactly when the mouldboard first appeared, but it was common on Chinese ploughs by the first century BC. This was many centuries ahead of the appearance of the mouldboard in Europe, where it arrived at the end of the medieval period.

Steering a plough

What is more, Chinese mouldboards were better designed than their European counterparts. In China the mouldboard was always made with the gentle, continuous curve familiar on all ploughs in use today. But in Europe mouldboards were made flat, and were much less effective. So while a European ploughman had to stop quite often to clean his share and mouldboard, because the latter did not throw the soil completely clear, his opposite number in China could go on ploughing without stopping for far longer, confident that the soil was being shifted continuously and evenly to one side of the furrow. This obviously also meant that the Chinese plough continued on its way smoothly through the soil with less friction than the Western version.

The northern Europeans had made some advances. As well as their flat mouldboards, they had added wheels to the plough. These made it somewhat easier to pull through the heavy soils of the region. But they still needed more than one strong ox or horse to pull the contraption. The Chinese farmer needed fewer oxen than the Western farmer, even when ploughing heavy soil.

Chinese ploughs had another refinement. There was an adjustable strut that connected the ploughshare to the horizontal beam of the plough. This allowed you to change the distance between the ploughshare and the beam, thus adjusting the ploughing depth. Such an adjustment was ideal when the farmer had to cope with different soil conditions – and since soil conditions are in a continuous state of variation, according to weather, seasons, and so on, this feature was invaluable. It was certainly better than varying the depth by leaning more heavily on the handle or beam.

The transfer of technology

So for much of the period known in the West as the Middle Ages, China was far ahead of the rest of the world in this technology. It was probably Dutch colonists and traders who brought back information about Chinese ploughs from Java in the seventeenth century. The Dutch East India Company was well established in Java, and had regular trade routes across the South China Sea to Saigon and Canton. Some time after this, a group of Swedes also visited Canton. It is known that they took away information about Chinese winnowing machines; they probably saw ploughs too.

The Dutch did not simply plagiarize the Chinese plough design. They adapted the mouldboard so that it would work well on Dutch soil. But the idea of a curved mouldboard persisted after the Dutch had shown how effective it could be. When Dutch engineers came to England in the late seventeenth century, they brought the curved mouldboard with them. It was quickly adopted in England, and took its part in the agricultural revolution that was soon to sweep the country. It also spread rapidly through other parts of northern Europe and America. The technology continued to be transferred. Thomas Jefferson, while Vice President of the USA, learned of an improved plough made in France. He applied his knowledge of calculus to the plough, thus demonstrating that it was made to the optimum shape.

How the plough evolved

Different soil conditions influenced the design of ploughs around the world. In Middle Eastern and Mediterranean areas, relatively lightweight ploughs are still used. In northern Europe, heavier ploughs evolved, together with other refinements to cope with heavy soils.

Roman ploughs

The Romans had a vast empire with land of many different soil types. In their southern lands they continued to use the ard, a strengthened version of the primitive scratch plough.

Later ploughs

The heavy soils of northern Europe made a heavy plough essential. Wooden mouldboards were also common by the medieval period. The weight made the plough difficult to drag through soil and large teams of oxen were required. Such a requirement must have reinforced the feudal system: the expensive beasts and plough would belong to the Lord of the Manor rather than individual farmers. Improvements came with the wheeled plough, which was easier to pull. More sophisticated designs of mouldboard and share, and better castings, also followed.

Draught beam

Handle

Stock

Egyptian plough, c.1800 BC

Early wooden plough, c.500 BC

Primitive ploughs

Some of our earliest evidence for what primitive ploughs were like comes from the wooden models made for burial in Ancient Egyptian tombs. These models reveal quite a simple plough, made up of a one-piece wooden share, a handle, and a draught beam to connect the plough to the harness. With ploughs like this, the kings of Egypt and Mesopotamia could produce the large food surpluses that were needed for their advanced civilizations.

Roman plough or ard, first century AD

Sole

Share

Chinese ox-drawn plough, c. AD 500

Oriental ploughs

China had one of the most advanced cultures when it came to the design of agricultural machinery. The first ploughs with iron shares appeared in China, and probably looked like the one shown here on a painted brick in a Chinese tomb. Another Chinese innovation was the mouldboard, to throw the soil to one side of the furrow. A mouldboard is just visible on this Chinese plough.

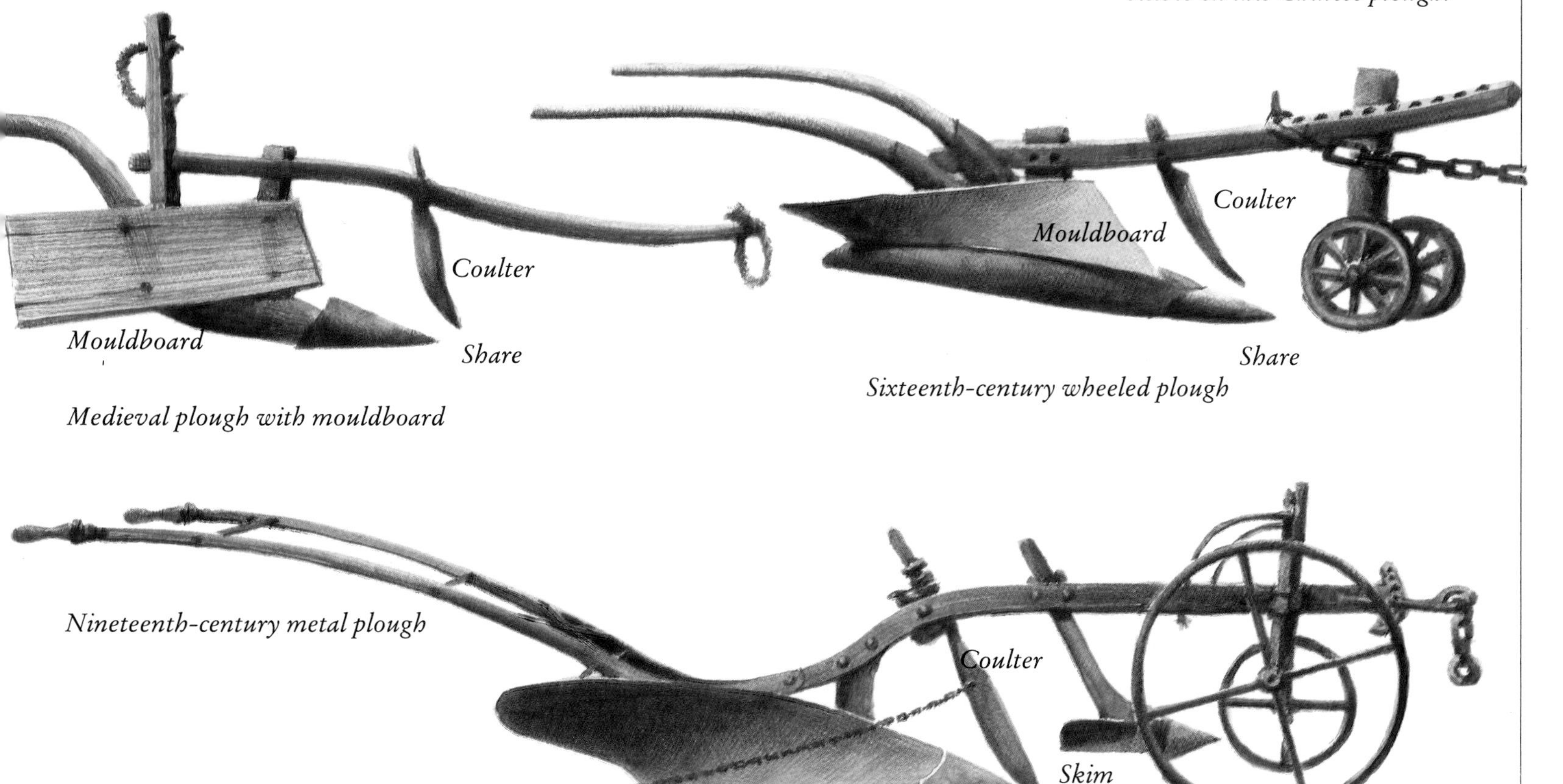

Medieval plough with mouldboard

Sixteenth-century wheeled plough

Nineteenth-century metal plough

SCIENTIFIC AGRICULTURE

Crop rotation, centring around the humble turnip, caused profound changes in agriculture

Inextricably tied to the land, agriculture is also tied to the prevailing social system. So any innovation that can improve agricultural techniques, making it possible to grow more food, or to grow it more easily, can be put into practice only if the social conditions are right. In medieval Europe, land ownership was organized according to the feudal system. A king would grant land to members of the nobility in return for certain services; nobles would grant land to their social inferiors in return for other services. And so on, down the line to the humblest worker, who would be allowed to cultivate a few strips of land in return for service (whether the provision of food, or military service in time of war) to his lord.

In such a situation, any change that involved a large outlay was very difficult. Even a simple plough and the animals to pull it would have to be shared. And an innovation that required working larger fields, rather than the strips of land cultivated by the medieval 'villein' (serf), would be impossible without a radical change in society. So unless there was some other pressure for alteration in the social system, innovation would tend to happen very slowly.

But even in the Middle Ages there were pressures that made it important to keep up agricultural productivity. The population was growing – albeit at a far slower pace than it did later – and these mouths had to be fed. One simple way to increase yield was to increase the use of the land.

Crops and rotations

The traditional practice in the Middle Ages was to alternate a season of cultivation with a season in which the land lay fallow because of the fear that the soil would be 'exhausted' by continuous cultivation. Farmers in many areas realized that they could achieve similar results, but greater productivity, with a three-year cycle, sowing wheat one year, perhaps barley in the second year and leaving the field fallow in the third year. Another alternative was to farm the field continuously for a number of years, before leaving it fallow for several more.

In each case farmers knew the benefit of the rotation, even though they did not realize the scientific rationale behind this benefit. In fact, what was happening was that the load of pathogens (disease-carrying agents) against one crop diminished when the farmer grew a different crop on the same ground.

Hoeing by hand

There came a point where some of the more adventurous farmers discovered that certain crops were more beneficial in a rotation than was a simple fallow period. In particular, turnips and legumes (plants such as clover, trefoil and lucerne) helped to increase yields while at the same time provide better animal fodder. In turn, this better fodder meant that higher quality manure was produced to put on the soil, increasing yields further.

It is uncertain exactly when this discovery was made. Historians used to talk about an 'agricultural revolution' that brought turnips and legumes to British fields quite rapidly during the eighteenth and early nineteenth centuries. One reason for this was provided by the agricultural activities of Lord Townshend, who left a career in politics in 1730 to farm his estate at Raynham, Norfolk. Townshend was an enthusiastic supporter of the turnip and of improved crop rotations. His work led to his nickname of 'Turnip Townshend'.

But Townshend was not the first to use these methods. Turnips had been grown in Norfolk for at least fifty years before 1730. He was not even the first publicist of the use of fodder crops. Writers of the late seventeenth century, and the novelist Daniel Defoe in 1720, had noted the benefits of this type of agriculture. On the other hand, the ideas of the agricultural reformers were not taken up instantly in the rest of the country, let alone the rest of Europe. East Anglia, especially the county of Norfolk, remained more advanced, although even here the new methods were adopted only slowly.

Nevertheless, Norfolk gave its name to one of the most successful methods, the Norfolk Four-course Rotation. This consisted of sowing wheat, turnips, barley and clover in successive years. By the 1830s and 1840s this rotation was becoming quite popular in England. In the period of recovery after the Napoleonic wars, and in the wake of the pandemic of potato blight, English farming began to develop once more.

One sign of this development was the establishment of several academic bodies connected with agriculture. The Rothamsted Experimental Station (founded 1843), the Royal College of Veterinary Surgeons (1844) and the Royal Agricul-

There was such great enthusiasm for the turnip in its role as 'agricultural improver' during the eighteenth century that its advocates might well have been accused of 'turnip worship'.

tural College (1845) all contributed to the movement to make agriculture more scientific. At the same time, more and more farmers were experimenting with the use of new fertilizers and foodstuffs. The introduction of cheap, machine-made land-drain pipes in the 1840s also helped the farmer get more out of the land.

But even such straightforward changes were related to a major shift in the social system. Efficient producers of fodder would want to own larger numbers of livestock. So the change in crop patterns goes along with a shift away from a straightforward feudal system to a society that included peasants who were rich enough to own both herds of cattle and sheds in which to keep them.

The coming of mechanization

Some of the best-known innovations in agriculture during this period were machines, devices that were designed to do basic agricultural jobs faster, more easily and more efficiently than the human labourers hitherto employed. The fame of the new agricultural machinery of the eighteenth century has encouraged people to compare the 'agricultural revolution' with the industrial revolution.

One of the most obvious, and most useful improvements in farm machinery was the adoption of improved ploughs. Designs for these filtered through from the Far East (*see* page 34). The plough was developed further by the addition of steam power in the nineteenth century, but this innovation was not used widely.

Devices for preparing the soil and sowing seeds, such as the horse-drawn hoe and seed drill, also appeared. Both go back to the seventeenth century in Europe, but were taken up on a large scale in the eighteenth century only after they had been popularized by the work of agricultural pioneer Jethro Tull. Even so, it was not until the mid-nineteenth century that these machines were used all over England. Most other countries, with the exception of The Netherlands, adopted them later still. A similar fate befell machines for threshing, chaff cutting and root slicing.

One area that seems, from a twentieth-century point of view, to be ripe for mechanization was the harvest. This was the most labour-intensive part of the crop-growing cycle, and agricultural pioneers spent much time devising mechanical reapers and machines for threshing (separating the wheat from the chaff). By the 1850s, mechanical reapers were quite common. One example was a bizarre machine designed by Patrick Bell, a Scottish Presbyterian minister. Bell's reaper

Bell's reaper
Patrick Bell produced the first practical reaping machine in 1826. At the front, triangular knives were mounted on horizontal bars, and a set of rotating sails pushed the corn between the knives. The drum at the back of the machine pushed the cut corn aside. Bell's machine worked, but it was not well made, and this, together with opposition to the very idea of harvesting machines in England, meant that it failed to catch on.

Machines for harvesting
The mid-nineteenth century saw a number of experiments in producing machines to speed up the harvest and reduce the amount of labour involved. Most of these machines were unsuccessful, but McCormick's successful design looked forward to the combine harvester of the twentieth century.

Hussey's reaper
Obed Hussey was the pioneer of the reaper in North America with a machine that came out in 1837. The cutting mechanism was positioned at the side of the reaper and the machine was pulled rather than pushed.

McCormick's reaper
Cyrus McCormick developed his reaper during the 1840s. It was a more refined design than its predecessors, and did particularly well in North America. There were two reasons for this. First, it was mass-produced, with interchangeable parts, making repair very easy. Second, farmers could pay for it by instalments, putting down a deposit of $35 and paying off the rest over 18 months.

was pushed rather than pulled by horses.

Bell's counterpart in North America was Obed Hussey, who designed a reaper that was pulled. This was superseded by another design, this time by Chicago-based manufacturer Cyrus McCormick. McCormick's reaper was shown in the London Great Exhibition of 1851, where it attracted much attention. Soon afterwards McCormick's factory was making 4,000 reapers a year.

However, even with this activity, much harvesting was still done by hand. The most important innovation for many farmers was the simple replacement of the sickle by the less back-breaking and more efficient scythe in the middle of the nineteenth century.

Threshing was first mechanized when the Scottish pioneer Andrew Meikle introduced his threshing machine in 1784. Threshing was a job that was hitherto carried out by hand, using a flail. It took about five days to thresh the wheat from one acre (0.4 hectares) using this method. Meikle's machine, which could be driven by water, horse or even steam power, could thresh the same amount of grain in under a single working day.

More forward-looking still was the combine harvester, a combined reaping and threshing machine that first appeared in the 1830s. This horse-drawn machine, ancestor of the modern harvester, cut the crop, separated the wheat from the chaff, and poured the grain into bags.

In other parts of the world different innovations transformed the harvest. In the USA in 1793, for example, Eli Whitney invented the cotton gin. This was a device that simplified the process of separating the cotton fibres from the seeds of the plant, previously a difficult, manual operation. The result was that farmers began to grow cotton on a large scale in North America for the first time. By the early years of the nineteenth century it had replaced tobacco as the main plantation crop in the southern states.

The seed drill

One machine that had the potential vastly to increase agricultural efficiency was the seed drill. The first farmers would sow their seed by 'broadcasting', that is to say they would walk through the field with a shoulderbag full of seed, flinging handfuls across the ground. Even with great skill and experience, it was impossible to sow seed evenly in this way. Some patches would be heavily sown; very little seed would fall in other areas. Sudden gusts of wind would further alter the pattern. The result was an uneven crop that would be wasteful because some plants, sown close

Jethro Tull's seed drill

together, would crowd each other out. It would also be more tiresome to harvest. And you would need to save a vast amount of the seeds from your crop (as much as half in some places) for sowing in the same wasteful way again next year.

The answer was to create a device that would deliver the seed evenly to the ground. The Ancient Mesopotamians, as long ago as 3500 BC, were probably the first to come up with some sort of solution: a primitive seed drill consisting of a container linked to a narrow tube that could be pointed down towards the soil. Seed could then be sown fairly evenly in rows, one row at a time.

Independently, and much later than the Mesopotamians, the Chinese created a more efficient version. This had several tubes so that you could sow a number of rows at once. Each tube also had a corresponding blade, like a little ploughshare, that cut a groove about three inches (eight centimetres) deep for the seed to drop into. Using a drill like this made sowing quicker, gave a more efficient use of the evenly sown seed, and meant that the yield was improved because the seed was put in at a uniform depth.

This type of Chinese seed drill was in use by the second century BC. It was used in India soon afterwards. Yet it did not spread to the West. When Western travellers began to travel to the Far East at the end of the Middle Ages, they heard of the drills. But the machines were mainly used in the north, far from the southern ports, and so few if any Westerners got to see them or understand how they worked.

However, the impetus for improvement was there. By the middle of the sixteenth century Western inventors were working on the problem, and a Venetian inventor patented a seed drill in 1566. Other designs followed, but none was efficient enough to be taken up. It was not until Jethro Tull's design of 1733 that the true advantages of an efficient seed drill were experienced in the West.

Jethro Tull was an outsider – a man who went into farming after working in the law. He was also a musician. His varied background gave him an original enough way of thinking to overcome the prejudices against new ideas harboured by many agriculturalists. He saw that the problem with previous designs of seed drill was that they were not able to deliver the seed evenly to the furrow, because there was no control of the seed flow. Jethro Tull solved this problem by adapting a mechanism from an organ. He used a brass cover and adjustable spring to regulate the flow of the seed.

It was a highly successful design, but even so it did not catch on immediately. It took enough people with the right manufacturing skills, plus a sales campaign by a drill maker in Suffolk, England, to convince people that the drill was worth the investment. But by the middle of the nineteenth century the use of drills based on Tull's design was common.

Stock breeding

Farmers had long known that it was possible to breed selectively, creating better varieties of sheep or cattle suited to particular environmental conditions.

In the eighteenth century further attention was given to the breeding of livestock. One of the greatest pioneers was Robert Bakewell, the specialist breeder who created the Leicestershire breed of sheep in the mid-eighteenth century. This was a breed that had a high proportion of flesh to bone, and could be fattened up quickly. Bakewell created it by inbreeding. He also laid heavy emphasis on improved animal care – better food, shelter and handling. His animals were more costly than their counterparts, and so their owners tended to care better for

Leicestershire sheep bred by Bakewell

them. If there was an improvement as a result of Bakewell's work it probably came from this care rather than from Bakewell's breeding methods.

Steam power and agriculture

During the nineteenth century, people started to save labour by using steam engines to power some farm machinery. By this time there were enough efficient steam engines small enough to be put on wheels and trundled out to the fields. Here they could power threshers in a fairly straightforward way, using a belt drive. But the powering of a moving implement, such as a plough, required more ingenuity. Pulleys and long cables were used to connect the plough to the stationary engine in Heathcote's steam plough, which appeared in the 1830s. As might be expected, the steam-powered thresher caught on widely, but many farmers were still using horses for ploughing in the mid-twentieth century.

Innovation and society

From a twentieth-century viewpoint, all the new agricultural developments have an obvious value. They were able to save the farmer enormous amounts of labour time. It seems odd that they were not taken up more quickly. But the situation in the eighteenth and nineteenth centuries was different. For one thing labour was cheap – there was less incentive to save on this than there is today. For another, machinery was costly. Most peasant farmers could not afford expensive seed drills or threshing machines.

So the innovators had to wait for their ideas to be taken up. With the enclosure movement bringing more land under cultivation in this period, farms were starting to become larger, and machinery gradually began to be more accessible. Meanwhile, it was often left to itinerant machine-owners to hawk their threshing machine and their skill around the country, offering the smaller farmers the chance to thresh their grain more quickly, for a fee.

Enclosure did away with the old open fields, as well as bringing under the plough land that had not been cultivated before. This replaced the old communal farms of the Middle Ages with farms structured much more like modern ones. This new system did not necessarily lead to more efficient farming, or to a greater take-up of machinery or other innovations. But it did make change easier.

Another set of social developments also probably had a key influence on agricultural innovation. As the industrial revolution gathered momentum, with the growth of cities and transport networks, so the demand in the towns for meat and dairy produce increased. This led more farmers to start growing fodder crops, and to switch to crop rotations like the Norfolk Four-course.

In America, innovations like the reaper and thresher helped people who were starting to farm the unsettled lands in the Great Plains area. But even here, Congress needed to encourage people to move to this dry region. By passing the Homestead Act in 1862, they allowed any settler 160 acres (65 hectares) of land if they lived on it and farmed it for five years. Additional land could be bought at a concessionary rate. It was these economic inducements, as much as the innovations, that pushed agricultural development along in the USA.

Historians who study the agricultural developments of the eighteenth century today are less eager than their predecessors to talk about an 'agricultural revolution'. The new ideas took a long time to catch on in many areas, and their take-up depended on a wide range of social factors. What is more, in many parts of the world – even in some countries that industrialized quite early – farming remained the preserve of manual labour that it had been for thousands of years.

BASIC SHELTER

How early humans came to grips with climate, environment and predators

The most popular image of the early human is of the 'cave man'. As the hominids gradually evolved into our species, *Homo sapiens*, and started to spread around the world, they encountered harsher climates than they had hitherto experienced. Caves provided shelter from the elements and a refuge from predators.

Of course, this is a highly simplified picture. We do not know for certain how humankind evolved or colonized the planet. And what is more, not all early humans could have lived in caves. There were no caves at many early settlements. And even where there is evidence of human settlement in caves, for example in the European painted caves, it is by no means certain that this is where people actually lived. The great decorated caves of France and Spain, with their painted chambers deep in the rock, were more likely to have been ceremonial centres than early homes.

But people did need shelter. And the evidence is that very early on in the history of mankind our ancestors learnt how to construct basic buildings to provide it. Of course, the evidence from the millennia before 12,000 BC is scanty. However, we can suppose that people used local materials and built simple structures. The people generally moved around a great deal, following hunting quarry such as mammoth, wild ox and deer, or travelling to the next place where they could gather vegetable food. So the first houses were probably temporary huts, or tents like the bison-hide tepees of the Plains Indians.

Tents made of animal hides on a wooden framework probably provided many early people with shelter. A tent has all the basic features required of a shelter: its sloping surfaces repel rainwater and stop the wind getting in; it also keeps the temperature even: warmer in winter, cooler in summer. And the tent's basic structure, a framework supporting a membrane, is echoed by the structures of other types of early home.

But the tent is not an ideal solution in many environments. It offers poor protection against predators, and is far from ideal in cold climates. Furthermore, it relies upon an adequate supply of large animal hides. People in many parts of the world soon began to experiment with more solid structures.

Wooden shelters

Where there were plenty of trees it was quite simple to make such a shelter. Tear down branches of more or less equal length (or use stone tools to cut them down). Make a framework of an inverted V-shape, binding the branches together with creepers, twigs or any other flexible material. Bury the ends of the branches in the earth, or weigh them down with stones. Cover the framework with leaves, skins or turf. The resulting tent-like structure would last well if necessary.

The main limitation of a house like this would be size. There might not be enough sizeable branches available locally to make a structure with enough height. Or if there were, they might be difficult to cut

Mud-brick settlement
The early towns of the Middle East were built of mudbrick. Access was often from above, via ladders.

Neolithic wattle and daub
A framework of spiit logs supports walls of wattle and daub and a thatched roof.

Stilt houses
A wooden structure supported on tall stilts was favoured in the swampy conditions of lake villages or in the tropics.

Skin tent
Amongst nomadic peoples a simple portable shelter of skins was often favoured.

Round tent
At Pincevent in France, tepee-like structure of poles supported coverings of skins.

Mammoth-bone huts

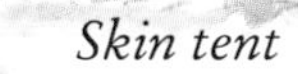

Mammoth-bone huts

Sometimes the early hunters' quarry provided material for a home. At several Asian sites, such as Mezhirich' in the Ukraine, Pushkari in Russia, and near Lake Baikal in Siberia, remains of huts built partly of huge mammoth bones have been found. A combination of branches, bones and skins seems to have been used to create different types of shelter. Some were like long, ridged tents and were probably covered with hides. Others seem to have been dome-shaped, with mammoth bones lashed to a wooden framework and supported on poles inside.

STONE AND CONCRETE CONSTRUCTION

Woodworking techniques were applied to stone with spectacular results, creating monuments to religion and royalty of astounding intricacy and beauty

Stone has been part of human technology almost from the beginning. The creation of the first stone tools was a key step in our development. But for thousands of years stone technology was quite small in scale. Hand-axes, knives and scrapers, querns for grinding corn into flour – these were the kinds of objects that our early ancestors made out of stone. Most of these items were probably made by the people who used them, when the need arose. If the hunt was successful, a new knife might well be fashioned there and then to do the butchering. It was a straightforward enough operation if your parents had trained you in the art of flintknapping, and the material was to hand.

To build with stone was an altogether more daunting prospect. The material would have to be quarried in large quantities; it would probably need to be transported some distance to the building site; and then it would need a number of people to construct even a very basic stone house. It is not surprising then, that, to begin with, only in places where one could not find materials like wood, which were easier to handle, did people venture to build stone houses.

However, this situation did not last for ever. Some early communities produced outstanding stone buildings at a very early date. For example, the people of stone-age Malta built the extraordinary stone temples at Tarxien around 3600 BC. Consisting of paired semi-circular chambers made of huge stone blocks, these temples would have required enormous effort, great skill, and a sizeable labour force in their construction.

But impressive as they are, the temples of Malta are isolated examples, the products of a culture that flowered rapidly and then mysteriously disappeared. Sooner or later a more sustained civilization would come along in which there was enough surplus labour to quarry, transport and work stone in large quantities, and in which there was an elite with enough power to organize and control such a labour force. The first outstanding example of such a culture was the kingdom of Ancient Egypt.

The pyramids

The civilization of Ancient Egypt was a strictly hierarchical one. At the top was the pharaoh, who ruled the two kingdoms of Upper and Lower Egypt. Next to the pharaoh in importance was an aristocracy of priests and scribes. Lower still were artisans and manual workers. Central to the pharaoh's power were the elaborate ceremonies devised by the priesthood, and the equally elaborate preparations made for the ruler's entry into the next world. Consequently, many of the buildings that have survived from Ancient Egypt are tombs, and the most famous of these are the pyramids.

The great step pyramid of the Pharaoh Zoser at Saqqara, and the mortuary complex that surrounds it make up the first large-scale monumental stone building that has come down to us. It was followed by a large number of other pyramids, mostly the more familiar-shaped true pyramids, including the most famous group at Giza. The pyramid at Saqqara dates back to *c.*2686 BC. The others are from later dates in the Old and Middle Kingdom periods, up to *c.*1777 BC. Together they are testimony to a society and a construction industry that were highly sophisticated. Planning, site preparation, quarrying, transport of the stone and the construction itself were all carried out in a highly effective way. The combination of refinement of technique and sheer muscle power that was required was awesome. How was it done?

The Egyptians have left very little written or pictorial evidence about how they built their great monuments. But by studying the buildings themselves, archaeologists have been able to put for-

ward some plausible theories. The first stages were planning and preparing the site. This could not be too far from the River Nile, since large quantities of stone had to be brought in, and the simplest way to do this was by water.

Once a site was found, it was first prepared by removing the sand and gravel so that the foundations of the pyramid would stand on bare rock. Then the area would have to be levelled, and this was probably done using water. An earthen bank would be built right around the site and the area within flooded. Then the builders would dig a series of channels in the rock, each with its bottom the same depth from the surface of the water. When the water was drained away, the spaces between the channels could be excavated to the same level, to give an almost perfectly flat surface.

Next, the exact positioning of the pyramid would have to be decided. Since pyramids were normally oriented according to the points of the compass, the master mason would have to make observations of the stars in order to find true north (the Egyptians had no magnetic compasses). The correct position for the base of the square pyramid would then be clear, and the mason could carry out the measurements to ensure that the area was indeed a square. Flax-fibre measuring cords, marked off in royal cubits (l royal cubit being about twenty inches. or fifty-two centimetres in length) were used. Since these could stretch, it is surprising how accurate the measurements could be. At the Great Pyramid at Giza, for example, there is a difference of only eight inches, twenty centimetres between the longest and shortest sides.

While all this was going on, quarry workers were already extracting the stone that would be needed for the pyramid. At Tura, on the eastern bank of the River Nile, limestone blocks were quarried. This would make up the outer casing of

the pyramid. Granite for the inner core was being extracted, probably at Aswan, down in the south of the kingdom.

Limestone, a soft rock that splits easily, was straightforward to quarry. The Egyptians had good copper chisels and sharp saws. They were able to chisel around three sides of a limestone block, then drive in wedges to make a split in the rock and thus remove the whole block. Sometimes they made the rock split by soaking the wedges and leaving them to expand – the expansion would then force a crack to open up.

Granite, far harder than limestone, must have been a much more difficult rock to extract. It would have blunted and bent the Egyptians' copper tools unless they had found a way of hardening them of which we are no longer aware. Perhaps they pounded the rock with a hard stone called dolerite, which can be found in the Eastern Desert. Whatever they did, it is unlikely that they were able to pick up enough granite in the form of loose boulders to fill the cores of the pyramids. Somehow they managed to remove the large pieces they needed.

Next, these great blocks of stone (weighing between 2.5 and 50 tons) had to be moved to the building site. Where possible, this was done by boat, but there was inevitably a land journey from the river to the foundations. To make this easier, the masons built a stone causeway from the Nile to the site. (This would also mark the route of the king's remains when his body was brought to its resting place.) It is almost certain that sledges bore the great stones on the land-bound parts of their journey. Modern experiments have shown that it is possible for as few as six men to drag a sledge carrying a six-ton block of stone.

Only now, after the planning, the site preparation and quarrying, and moving the stone, could the actual construction begin. We know very little about how the building was done. The central question is how, without the machinery of the modern contractor, did the Egyptians raise the enormous stone blocks to the required height? The most widely held belief is that ramps were constructed to allow the stone to be raised. These might have been helical, working around the pyramid in a sort of square spiral. Or there could have been a long single ramp at right-angles to one side of the pyramid.

Both types of ramp would have had to grow with the pyramid itself. They would also have left a great deal of masonry to be cleared away at the end, little of which has survived. Nevertheless, some form of ramp was almost certainly used, and it is testimony once again to the huge and well-organized labour force that there remain so few traces of these large, temporary structures.

The pyramids are worth dwelling on at length because they show how far civilization had come from the first primitive shelters that early people built to protect themselves and keep themselves warm. Unlike these early structures, the pyramids required careful planning, well-developed skills in handling materials, a large force of skilled and unskilled labour, an effective transport system, a highly efficient administrative hierarchy to supervise every aspect of the construction, and enough of a food surplus to supply the army of workers.

And just as the creative process was different, so was the result. In contrast to the ad hoc structures of early people, the Egyptians could create buildings that were large, expensive, and capable of conveying some sort of cultural message – in the case of the pyramids, the greatness of the Pharaoh and his survival into the afterlife. After this, buildings would never be the same.

Other civilizations, much later in time, but similarly reliant on very basic technology, produced impressive stone buildings. Among these are the central American pyramids and temples of the Mayan peoples and the fortresses of the Incas in South America. The latter contain blocks of stone weighing more than 100 tons.

The progress of construction

The earliest stone builders did not join pieces of stone together using the methods that are used today. True, the Egyptians often used thin lime mortar, but this was as much used to lubricate the massive stone blocks as they were pushed into position than as a cement to keep them together. In the pyramids the joins are so fine that the blocks stay together under their own weight.

Other early stone builders, accustomed to wooden construction, used carpentry methods to join their stones. A famous example of this technique is in the joints of the enormous blocks that make up the ancient British monument of Stonehenge. Here the uprights are fastened to the lintels by means of mortice and tenon joints – a protrusion in one block neatly fits inside a hole in another. This is a technique that is still used by woodworkers today.

Like the Egyptians, the Ancient Greeks also cut their stones very accurately in their monumental buildings. But they used an additional means to keep the pieces together; iron rods and cramps held

Greek temples
The structure of typical Ancient Greek temples, such as the Parthenon, Athens, is based on a forest of pillars holding up massive lintels topped originally by tiled roofs. The individual stones of the pillars are held together with iron rods, like the wooden dowels carpenters sometimes use.

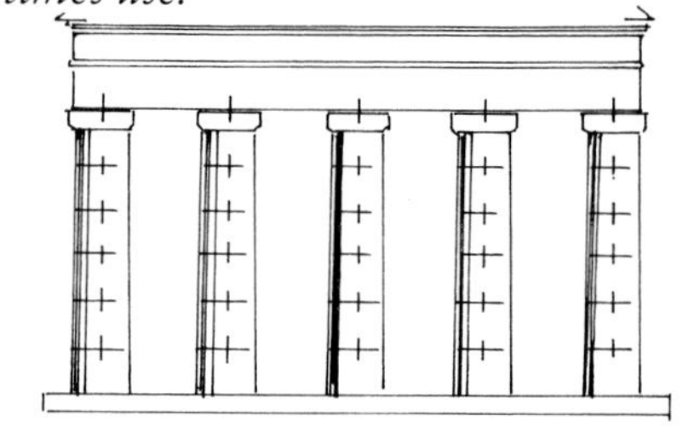

Stonehenge
In this famous British monument of the Bronze Age, the great circle of sarsen stones with their massive lintels are joined together with mortice and tenon joints, another carpenter's technique. The projecting tenons on the uprights and the mortices in the lintels are clear in a cross-section of the stones.

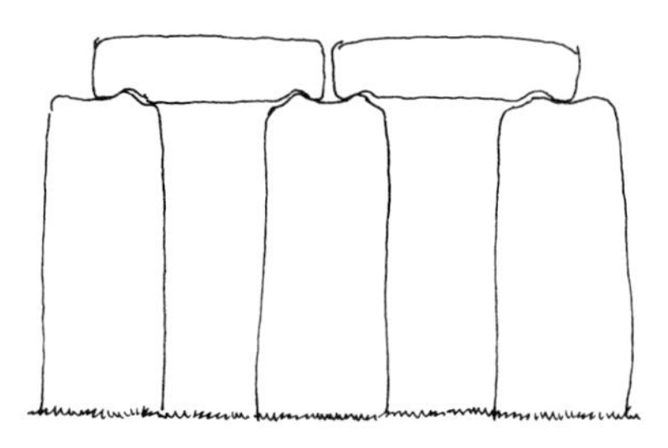

Inca buildings
The Inca city of Machu Picchu in Peru has many stone buildings. These reveal stone-cutting of the utmost skill – the individual blocks are shaped so precisely that the builders could fit them together without mortar.

together the stones of many a Greek temple. This was also a method used by the Romans in fixing the stone outer cladding to the facades of great structures like the Colosseum. But they did not use iron in the massive foundations required for buildings like this. For this task they had a material that was just as strong, but much easier to use on a large scale: concrete.

The Roman contribution

In fact, the Romans used a number of different types of masonry for different purposes in their buildings. One of the simplest styles was what they called 'opus quadratum', finely cut rectangular blocks of stone held together either with mortar or, in the Greek style, with dowels or cramps. To begin with, walls of opus quadratum were made of very large blocks of stone, usually 4 x 2 x 2 Roman feet (1 Roman foot = 11 .7 inches or 29.6 centimetres). But as mortar began to be improved in the third century BC, and as builders started to realize how strong it could be, smaller stones began to be employed. These were often irregular in shape, and were known as 'opus incertum'.

A more regular version of this, with the stones laid in a diagonal pattern, was called 'opus reticulatum', and there was a still later and yet more regular style, using brick facing, 'opus testaceum'.

Incertum, reticulatum and testaceum types were all used as facing to walls of concrete, the use of which was one of the most important Roman contributions to architecture. Roman concrete consisted of fragments of stone or brick mixed in a mortar of lime and sand. The stone could be one of a number of Roman building stones tufa, travertine, or peperino. The sand was usually a volcanic earth called pozzolana. Together these substances made a material that was strong and easy to work. The use of pozzolana meant that the concrete would set under water, making it invaluable for bridges and harbour walls. It allowed the Romans to explore on a large scale architectural features and techniques, from very heavy foundations to vaults and domes, which had only rarely been attempted before.

Concrete allowed the Romans to use unskilled and semi-skilled labour in building in ways that were not possible before. The construction of a massive concrete foundation, for example, was very straightforward. A wooden form of posts and planks would be built. Alternate layers of mortar and stones would be poured in and allowed to set. The wooden form would then be removed. A very similar method is still used in concrete buildings today indeed, on many modern buildings one can see the impression of the grain of the wooden 'mould' that kept the concrete in place.

Once the foundations were in position, the rest of the construction could be built on top. A similar method was used, except that outer walls of stone or brick were substituted for the wooden form. When these were built up to a certain point, the concrete core could be poured in as before, producing a wall of great strength and thickness. Skilled workers were required only for the outer 'skins'.

After the Romans

The Byzantine empire inherited many of the Romans' methods of construction. They used concrete as a mortar in many magnificent buildings, such as Justinian's great church of Hagia Sophia in Constantinople (now Istanbul). In the West, however, the secret of making concrete that would set under water was lost in the Middle Ages. One person who made a useful substitute for the Roman's pozzolana-based concrete was the sixteenth-century French architect Philibert de l'Orme, who suggested a concrete made of quicklime, river sand, pebbles and gravel for harbour walls and similar constructions.

But it was even later that the art of making true 'Roman concrete' was revived. This was in 1756, when British engineer John Smeaton built the third Eddystone Lighthouse (in Devon) on foundations of concrete made with pozzolana brought from Italy. The next major deveopment occurred when people began to strengthen concrete by reinforcing it with iron rods. But that technique belongs to a different era of building.

The great buildings of Egypt, Greece and Rome became symbols of the power of the people who ordered them to be built, of the importance of the state, or of the glory of a god. Ornamentation was a good way of pointing up the symbolic power of a building, and stone was a good medium for carving. So great ages of stone building have often also been great ages of architectural ornament. The Ancient Greeks exemplified this. They evolved the system of the three 'orders'(Doric, Ionic and Corinthian), which laid down the standard way of designing and ornamenting each column, base and capital, and the entablature with its frieze above the columns.

VAULTS AND DOMES

Architectural flights of fancy led to the development of the buttress and opened up new possibilities in building

Broadly speaking, the architectural technologies described so far in this book could be used to create two types of structure – wall buildings and frame buildings. In a wall building, a solid wall is the key element in the structure that holds the building together and bears the weight of the roof. In a frame building this role is played by a network of vertical and horizontal posts, which can be filled with non-load-bearing walls, windows or partitions. A Mediterranean mud-brick house, with rectangular walls holding up a flat roof, is a good example of a wall building. A north-European half-timbered house, with the weight taken on a wooden framework filled in with wattle and daub, is an example of a frame building. Today, many conventional brick-built houses are wall buildings, whereas most tall buildings, such as skyscrapers, are frame structures.

The two types of structure could handle the problem of roofing in different but related ways. In the Mediterranean mud-brick house, the roof often made use of a separate flat-frame structure attached to the tops of the walls. In the north-European half-timbered house, roof and walls were often indistinguishable, the whole house being based on frames shaped like an inverted letter 'V'. Both these types of building, then, used frame technology for their roofing.

Early solid roofs

There is another solution to the question of how to roof a building, which is particularly appropriate when using brick or stone. The method in question is called corbelling. If, on reaching the top course of the outer walls, the builder lays another course of stones that slightly overlaps the interior, and if this is followed by further courses similarly overlapping, the courses will eventually meet in the middle to form a sloping roof.

If the overlaps are not too great, and the span is not too large in relation to the weight of the stones, such a corbelled roof can be both strong and durable. If the building is rectangular, it is in effect a primitive vault; if the building is circular, it is a primitive dome.

Roofs like this are known to have been produced in quite early, primitive cultures. Where wood for frame roofs was scarce they could be a necessity. And they constituted some of the most beautiful early structures. Particularly noteworthy examples are the domed 'beehive houses' of Ancient Cyprus, which were built over 5,000 years ago.

'Treasury of Atreus', Mycenae

The same technology was being used in the Mediterranean during the time of the Mycenaeans, those early Greeks who flourished around 1300 BC. Some of the Mycenaeans' most famous surviving buildings are subterranean domed tombs, the best-known example being the so-called 'Treasury of Atreus'. This building consists of an imposing corbelled pointed dome some forty-eight feet (14.6 metres) in diameter, and almost as tall inside. It is made of thirty-four courses of stones, each one cut to form a curved surface, and fitted exactly to its neighbours. There is a single cap-stone on top. The whole dome is covered by an earth mound and so cannot be seen from inside, but it creates an imposing interior.

The Roman contribution

Buildings like the Mycenaean tombs show how sophisticated corbelled construction could be. But it was not until over a thousand years later that the next big advance in solid roofing came about; this was made by the Romans. They inherited their architectural tradition not from the Mycenaeans, but from Classical Greece, the civilization that flourished in the fifth century BC. These Greeks had begun to build very large public buildings – including their famous temples such as the Parthenon at Athens. These were buildings that had a simple pillar-and-post construction. The wider the roof span, the more rows of columns needed around the edge to support the horizontal beam of wood for the roof.

But the Romans were also inheritors of another tradition, the architecture of the Etruscans, who had preceded them in Italy. The Etruscans used the stone arch, and the Romans seized on this for construction. It gave them supremacy in bridge building, with structures like the Pons Fabricius in Rome. When combined with Roman concrete, it also gave the opportunity for a different approach to roofing.

A rectangular room with an arch at either end lent itself well to the construction of a simple, semi-cylindrical barrel vault. Using the Romans' strong concrete, making such a vault was easy. You simply had to build the walls and arches, construct a wooden mould (the 'centering') in the shape you required the vault to be, and pour in the concrete. When the mixture had set hard, you removed the centering

The Pantheon, Rome, was rebuilt by the Emperor Hadrian in AD 110. Its coffered ceiling lends the masonry both lighter weight and greater beauty.

0
10
20
30 meters

to reveal the vault. This gave you a roof that was quick and easy to make without using highly skilled labour.

The barrel vault could also be adapted to fit different sizes and shapes of building. If the span was large, the concrete could be poured into a framework of brick ribs. These ribs reduced the risk of the concrete cracking, and also meant that the centering did not have to be so heavy and cumbersome.

A square space, or the intersection of two barrel vaults over a square area, could be treated in a slightly different fashion. Two vaults could intersect above the square in a diagonal cross-shape; the lines of intersection would then appear as 'groins' on the surface of the vault. The same device could be used to roof a long building divided up by pillars into a number of square bays. These vaulting techniques typified the classic Roman public building, the *basilica,* which acted as meeting place, governmental chamber and law court.

Concrete was also ideal for building domes, another skill at which the Romans excelled. One of their best, and best-preserved buildings, the great circular temple in Rome known as the Pantheon, is roofed with a vast concrete dome. The great span, extending some 142.5 ft (43.4m) is the same size as the diameter of the building. It is bridged by a dome that is several feet thick, and which rests on a wall of concrete faced with brick that is constructed to take all the strains from above. The weight of concrete is reduced by leaving out the centre of the roof – a circular 'eye' at the centre of the dome is open to the sky.

And domes, especially large ones like that at the Pantheon, do create difficult stresses and strains. The Romans knew that the art of dome building was not simply a matter of building a large hemispherical concrete shell. They needed to devise buttresses of various types to help the walls take the strain and pass it to the ground. Again, the Pantheon demonstrates these devices. The walls are lined with constructional niches topped with small semi-domes. There are also spur buttresses that help to direct the weight of the dome towards the ground. Both types of device are concealed within the structure of the wall.

Further strengthening devices, such as a stepped thickening of the dome at its lower levels, help to reduce the risk of cracking. The square, coffered effect on the interior of this and other domes also helps to make the structure seem lighter. The concrete is thinner in the central squares of the coffered pattern than it is on the intersecting ribs.

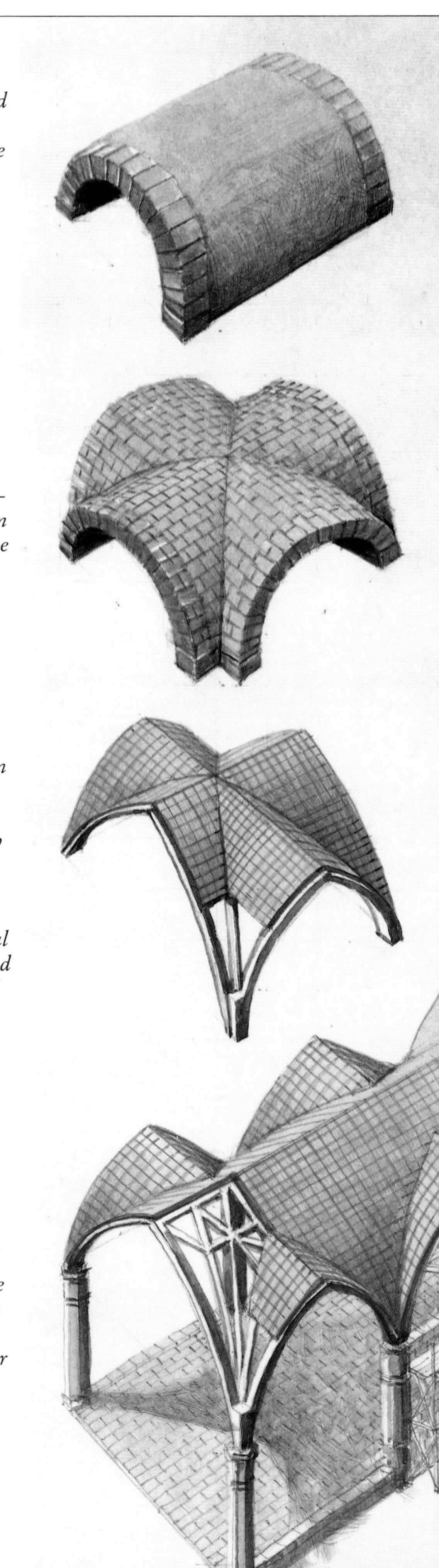

Barrel vault

The simplest form of vault is the barrel vault. This is semi-cylindrical in form, and is made up of a succession of semicircular arches placed one after the other along the length of a building. It is adequate for roofing long, narrow rectangular buildings, but wider buildings are a problem because the height of the vault has to be increased in proportion to the width. Barrel vaults were used by the Romans and later builders. Concrete was obviously useful to create a smooth finish.

Intersecting vault

Where two barrel vaults joined at right-angles, this type of intersecting vault was produced. The surfaces of the adjacent vaults would come together to form an X-shaped cross. Such a vault was useful at an intersection point in a building, such as the crossing in a church. It could also be used to roof any square section of a building, and a row of intersecting vaults could be lined up to roof a rectangle in a series of cells or 'bays'.

Gothic ribbed vault

Barrel vaults were heavy, and needed strong wooden 'centering' to support them while they were being built. They also required very thick supporting walls to take their thrust. So masons came up with the idea of the ribbed vault, in which the intersections are marked by strong stone arches or 'ribs', which take some of the strain. The Gothic masons of the medieval period also used pointed rather than round arches. These made it much easier to roof rectangular, rather than square bays.

More ribs

As time went on, the Gothic masons increased the number of ribs in their vaults. This gave a yet stronger framework to support the stonework and channel the stresses and strains. It had the added advantage of producing intricately patterned ceilings, which became one of the glories of medieval churches and other public buildings.

Gothic architecture
The European masons of the high Middle Ages built in the Gothic style. The most important feature of this style was the pointed arch. This was a great improvement over the round arch because it was possible to adjust the relationship between the height of the building and the width of the arch with ease. It also gave Gothic buildings their characteristic 'upward soaring' appearance.

Flying buttresses
Another feature of Gothic architecture is the flying buttress, the outside support that takes the strain of the heavy vault down the outside of the building.

Outer roofs
The stone vault was usually topped by a pointed roof supported by a wooden framework.

The building community
A large medieval building needed a sizeable workforce. Their workshops would be in temporary timber structures on or near the building site.

Trusses – the power of the diagonal

The Romans were also the first people to make widespread use of trussed wooden roofs. In this type of design, extra strength is achieved by the use of diagonal cross-members. The Greeks had begun to build such roofs as a means of covering broader spans in their temples. However, it was the Roman architect and writer Vitruvius who publicized the idea. One of his buildings, at Fano in northern Italy, used a trussed roof to cover a space 120 x 60 feet (36.6 x 18.3 metres).

Domes over a square space

The dome of the Pantheon is a magnificent achievement. It was reproduced with variations, and on a smaller scale, in numerous buildings across the Roman empire. But the Romans did not manage to build a dome over a square space. This challenge was left to the engineers of the eastern, Byzantine empire that followed Rome.

The problem was to discover a way of bridging the gaps (roughly triangular) occurring between the outline of the circle and the corners formed by the walls. Various devices were tried, but with slight success. The simplest solution was to place a stone slab across each corner to form an octagon. This was rather inelegant, and was restricted to smaller spans. Subsequently, architects experimented with series of small arches, or corbelled masonry, above the corners. These met with some success. However, it was the Byzantines who came up with the solution – the pendentive. This is a curved overhanging surface, again roughly triangular, that makes the successful transition between the square and the circle.

Once the Byzantines had developed the pendentive, they were then able to use the dome with the same freedom with which the Romans had used the vault. The dome was no longer limited to round, or even square buildings. Rectangular spaces could be divided into bays, and then roofed with a series of domes on pendentives. The Byzantines also developed the use of semi-domes around the edges of square, domed buildings. These had the twin advantages of giving more flexibility to the shape you could roof with domes, and of being useful devices for carrying stresses in heavy structures.

Gothic vaulting

During the Middle Ages, master masons in the West took vaulting in a radically different direction. They were at an advantage in that they discovered the usefulness of the pointed arch, the hallmark of Gothic architecture. With a typical Roman semicircular arch, the architect is limited to using cross vaults over square spaces, either singly, or in a series to vault a rectangular space.

Another solution to vaulting rectangular spaces is to raise the arches on the smaller sides on taller pillars, but this is not always very effective visually, as it results in the arches looking rather ungainly, as if they are placed on stilts. However, using the pointed arch, rectangles of all shapes and sizes can be vaulted easily, because with this type of arch it is possible to vary the relationship between the diameter and the height of the arch.

Squaring the circle

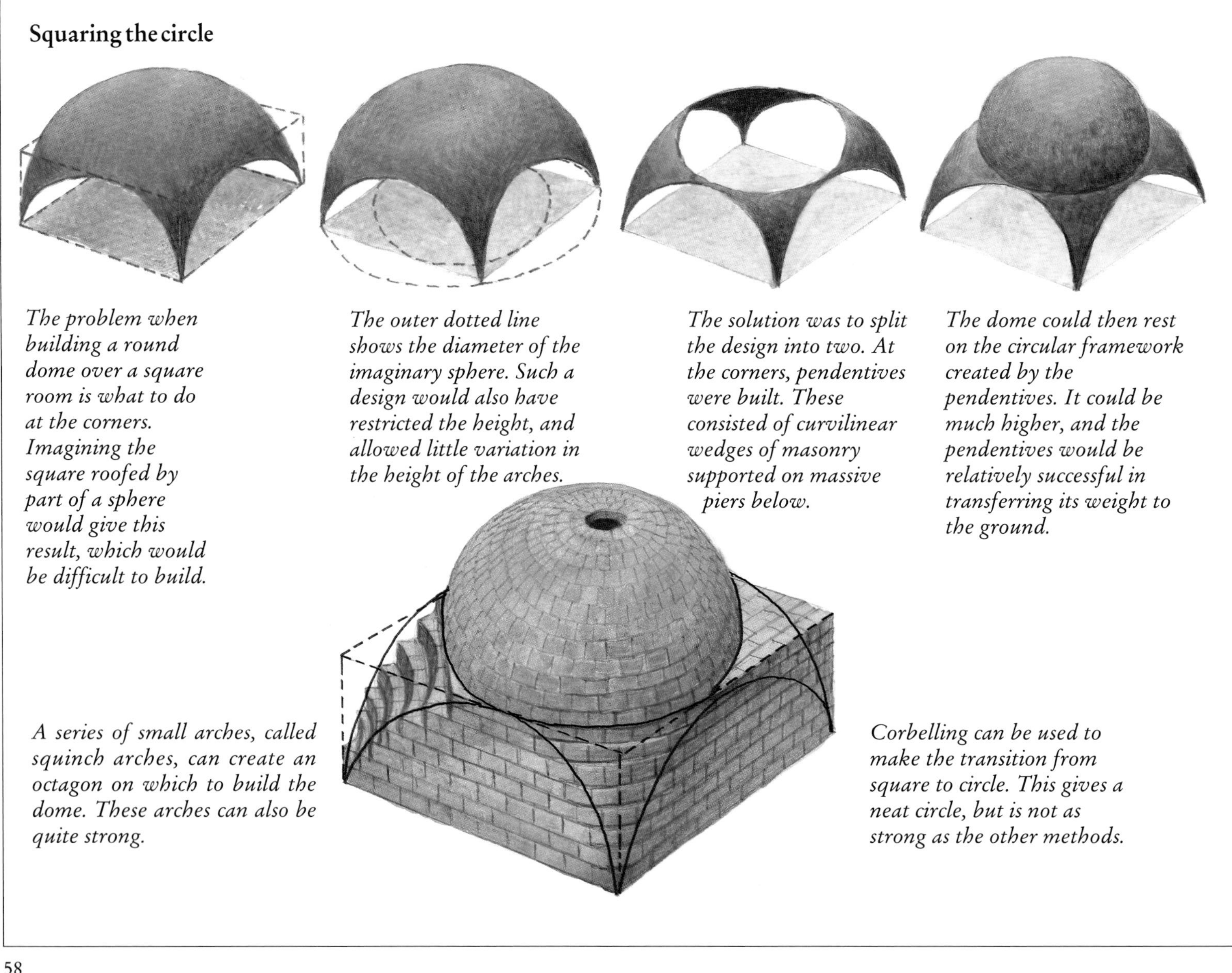

The problem when building a round dome over a square room is what to do at the corners. Imagining the square roofed by part of a sphere would give this result, which would be difficult to build.

The outer dotted line shows the diameter of the imaginary sphere. Such a design would also have restricted the height, and allowed little variation in the height of the arches.

The solution was to split the design into two. At the corners, pendentives were built. These consisted of curvilinear wedges of masonry supported on massive piers below.

The dome could then rest on the circular framework created by the pendentives. It could be much higher, and the pendentives would be relatively successful in transferring its weight to the ground.

A series of small arches, called squinch arches, can create an octagon on which to build the dome. These arches can also be quite strong.

Corbelling can be used to make the transition from square to circle. This gives a neat circle, but is not as strong as the other methods.

Bridges – another type of span

Bridges followed a development in some ways similar to roof spans. The earliest bridges to survive today are simple pillar-and-post structures that date from prehistoric times. Flat, stone slabs bridge the gaps between stone piers set at intervals across a river. Such a design is possible only when the gap is narrow enough and the river shallow enough to allow the piers to be built. A major breakthrough came when it was realized that the arch could be used in bridges. Simple semicircular arches could cut down the number of piers and provide a structurally stable design. The use of a shallow segmental arch was even more beneficial because you could span a wide gap without going too high. The Chinese were the first to use this idea, in the Great Stone Bridge over the River Chiao Shui, designed by Li Ch'un in AD *610. It has a span of some 123ft (37.5m), around twice the span that the Romans achieved with bridges using semicircular arches. The design of the Great Stone Bridge looks strikingly modern: it is the ancestor of many of today's concrete bridges.*

Working in stone rather than concrete, the Gothic masons also evolved a different method of vault construction. Rather than having the intersections meet as groins, the medieval vault is based on a network of stone 'ribs' that both mark the intersections of the panels, and also form the basis of the structure. They support the thin panels, which were inserted on top of them.

This was an ingenious solution, and it resulted in some of the most beautiful ceilings ever constructed, the networks of ribs and carved stone bosses where they join forming a fitting climax to the interior design of many a fine cathedral. There was also a major constructional advantage, because the supportive strength of the ribs meant that the masons no longer needed the heavy centering that the Romans had had to use with their concrete vaults. With Gothic cathedrals getting higher and higher as the Middle Ages went on, this was an increasingly important benefit.

However, this type of vault also posed constructional problems. For a start, many of the ribs join at the four corners of the vault, at the top of the supporting pillars or piers. This meant that the latter had to be specifically designed to support the vault, and yet still to join it in a visually satisfying way.

Another problem was the perennial one of the weight of the vault. This seemed to conflict with the desire of the cathedral-builder to include windows that were as large as possible – walls of glass could not support stone vaults. The answer was to develop more sophisticated buttresses. Spur buttresses, like those the Romans had used, were employed. But Gothic masons also used flying buttresses, basically an arch starting at ground level as a pier totally detached from the building, and then rising to meet the wall and take the weight of the vault.

Working out the stresses and strains involved in complex structures like this was not easy, and a medieval master mason had to be something of a mathmetician. Not surprisingly, there were mistakes. The great arches at the crossing of Wells Cathedral, England, for example, had to be strengthened with extra arches.

Once these structural principles had been grasped, Gothic vaulting developed, in the sense that there was more and more elaboration in the patterns made by the ribs. Designs that had begun as functional became more and more decorative as masons realized the scope that the ribs gave them. But the vaulting kept the same structural basis, to which the flying buttress was the key.

The Renaissance, which saw the eclipse of the Gothic style and a return to styles influenced by classical Greece and Rome, produced many of the world's greatest buildings. Nevertheless, there were few fundamental structural innovations. Substantial progress in building construction had to wait until the industrial period, when materials, especially metals, were to be used in new ways, and when concrete would make new, dramatic appearances on the buildings of the world.

METAL IN BUILDING CONSTRUCTION

Iron proved its usefulness in bridges, open-plan factories and high-rise buildings, and led to the invention of the elevator

Britain was the first country to experience an industrial revolution. Its effects began to be felt in the eighteenth century, and from the point of view of the construction trades, the first widespread changes were to do with providing improved transportation. The canals of the eighteenth century were followed by the roads, bridges and railways of the nineteenth. Important new technologies were employed in these construction projects. The most revolutionary of these technologies, from the point of view of the building industry, was the use of iron, with its considerable advantages over wood of resistance to fire and decay.

The most famous, and most startling example of this new trend was the first bridge made completely of iron, at Coalbrookdale in the west of England, at the place that has since been named for this feat of engineering: Ironbridge. It was built in 1779 by the iron-founder Abraham Darby III, probably to designs by the architect T. F. Pritchard, who died two years before it was completed. In some ways it is not an outstandingly original design – it follows the pattern set by the many arched bridges that had been built in wood and stone during the preceding centuries. But it showed once and for all the enormous potential of this 'new' material for construction.

Another work of engineering, the Pont Cysyllte Aqueduct in North Wales, was similarly influential. Built between 1795 and 1805 by another well-known figure of the Industrial Revolution, the engineer Thomas Telford, it carries a cast-iron trough across a series of high stone arches over the River Dee. The trough contains the Ellesmere Canal. Once again, it stood as a large and very public advertisement to the strength and versatility of iron.

Such bridges inspired designs of still more daring. Telford returned to an all-iron design in 1801 when he submitted his proposal for rebuilding London Bridge. This scheme, which was not carried out, involved a single, great and graceful iron segmental arch across the River Thames. Its span was to have been 600 feet (183 metres) and the bridge was to have risen some 66 feet (20 metres) above the water level at the centre. Although never built, the bridge was technologically possible – and it was another project that fired the imagination of all who saw the drawings.

At around the same time as the engineers were creating structures like these, architects were starting to experiment with the use of iron in buildings. James Wyatt used iron in the structure of a new palace built for King George III at Kew, London, between 1802 and 1811. This building has since been demolished. But a more famous one, the Royal Pavilion at Brighton, has survived. John Nash's fantasy house for the Prince Regent contains iron staircases and iron columns – the latter disguised as the trunks of huge palm trees – that hold up the roofs above such rooms as the great kitchen. Other designers began to experiment with iron supports for balconies in churches, although these were criticized by architectural purists as wholly inappropriate for church interiors.

Mills, factories and warehouses

By this time builders were already experimenting with iron in the structures of buildings such as factories and warehouses, where the new material did seem much more in keeping. It was soon realized that a system of cast-iron pillars and beams could form an ideal skeleton for an industrial building. It could provide interior spaces with large, open floor areas, ideal for housing large machines or storing vast quantities of produce. It also allowed for a greater number of storeys than a conventional design. A spinning mill at Shrewsbury owned by Benyon, Bage and Marshall, and built in the 1790s, was the first to use a structure completely made up of iron columns and beams. Soon, many other warehouse- and mill-owners followed suit, as they saw the advantages. The pattern of the industrial building in the West – a design made up of regular bays and based on an iron frame – was set for many years to come.

Traditional materials still had their place in these buildings. The outsides were usually clad in brick, and shallow brick vaults were often used between the iron beams beneath the floors. Using brick in this way helped to fireproof the buildings. Some, such as the imposing buildings around the Albert Dock in Liverpool, even had classical columns supporting the weight of their structures. However, closer inspection reveals that even these were made of cast iron. Nineteenth-century factories frequently looked rather more conventional than they really were.

And there certainly were some new and unconventional ideas behind the designs of some of these early factories. In one such mill, built in 1801, steam from the company's Watt engine was piped through the cast-iron columns to heat the building. As the century went on, factories began to acquire gas lighting.

Indeed James Watt, the man who made the steam engine efficient and, with his partner Matthew Boulton, became a successful manufacturer, was well placed to enter the world of cast-iron architecture. He had the inventiveness of mind to grasp the requirements of a new method of construction, and he could make the components in his factory. It is not surprising, then, that he was involved in the design and construction of one of the first iron-

Mines yielding coal and iron ore were at first serviced by horse-drawn railways. From such humble beginnings came technological achievements like the first iron bridge, so famous that the town where it stands is still called Ironbridge.

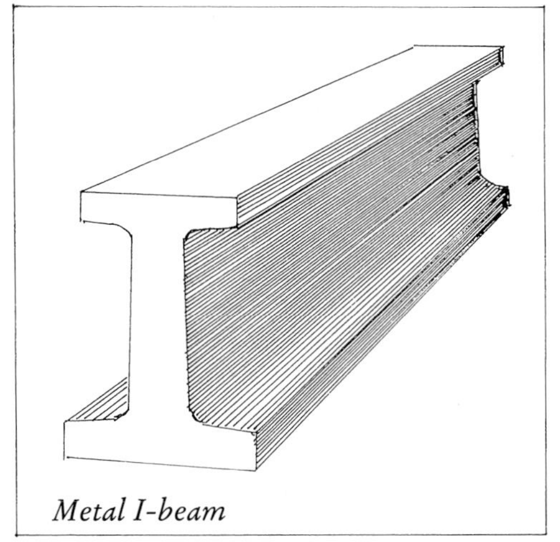
Metal I-beam

framed buildings, Philips and Lee's cotton mill in Salford, Manchester (1801).

One of the most important innovations in this cotton mill was the use of the I-beam. This type of metal girder has been used ever since in the frames of buildings, and engineers now recognize that its combination of strength, rigidity and relatively light weight is the best – as yet – that they can achieve. Watt's mill was also innovative in another way. It had seven storeys, pointing the way to still higher metal-framed buildings. In the industrial north of England, high-rise building was beginning.

Other men from backgrounds outside architecture contributed to building technology in the early nineteenth century. The American railway-builder Robert Stevens developed a technique for making rolled-iron railway tracks that was later adapted for the production of beams for buildings. And Manchester engineer and shipbuilder William Fairbairn designed buildings in which the floors were separated by a layer of concrete, poured on to thin iron plates set between the girders. In this way, he achieved effective fireproofing without the brick arches most commonly used in buildings of this type. Both of these developments started to be taken up in the 1840s.

Prefabrication

All the buildings described so far had brick or stone outside walls that provided often elegant façades and also played their part in supporting floors and roofs. The next step was to take the metal skeleton to the outside edge of the building, so that it could bear the whole of the weight. The outer walls could then be made of relatively lightweight in-filling material – metal or glass sheets. This concept came from America, particularly from the pioneer builder James Bogardus.

Like the great eighteenth-century British ironmaster John Wilkinson, Bogardus had unswerving faith in the 'new' material. For Wilkinson, this meant making everything, from machine tools to coffins, out of iron. For Bogardus, it meant designing every kind of metal-framed building, from houses, through shops and factories, to a vast tower and amphitheatre for the 1853 New York World's Fair. Not all of these buildings were actually constructed, but in the large number that were, Bogardus picked up on yet another trend of nineteenth-century industrialism: mass production. By using parts of the same size and specification, he could put up similar buildings all over the USA: only the number of storeys and bays need vary, according to the needs and purse of the client, and the size of the site. Soon, Bogardus was not the only builder working in this way in the USA – in the second half of the nineteenth century almost every sizeable town had shops, factories, or warehouses built using these methods. These buildings often had tall, regular façades made up of a grid of cast-iron beams and columns often filled with windows or with sheet metal. Large, open floors regularly punctuated with pillars typified the interiors.

Getting to the top

Bogardus was among the first to realize that high-rise buildings needed some mechanical means of getting people to the upper floors. In his unbuilt project for the New York World's Fair he suggested a steam-powered hoist to take people up the tall tower. Of course, mechanical hoists had been used before in factories, but these were not considered safe for people – if the rope broke you would plunge straight to the bottom of the shaft and at the very least be seriously injured.

The breakthrough was not provided by Bogardus, but by Elisha Graves Otis, whose 'safe elevator' was first seen at the New York World's Fair of 1853. Otis demonstrated his device by asking for the rope to be severed while he was travelling in the elevator. It was brought to a stop immediately by ratchets fitted to the girders in the shaft. 'All safe, gentlemen!' said Otis, making history with an epigram, and attracting orders, too. In the late 1850s elevators began to be installed in New York department stores. Soon they would spread to Europe.

During the mid-nineteenth century,

Elevators

Elisha Graves Otis was an experienced mechanic who had seen and worked in many factories containing hoists for goods. These were notoriously unreliable. They often had a counterweight on the end of a hemp rope, and these ropes would break with depressing regularity. So he came up with a simple safety device. He installed vertical ratchets to the upright girders in the elevator shaft. In the event of an accident a spring would be activated. This sent out a stay that caught on the ratchet. The elevator would be brought to an almost immediate stop. This was the principle behind Otis's first 'safe elevator' for passengers, originally demonstrated at the New York World's Fair of 1853. The idea soon caught on, and developments in motors and drive systems made Otis's elevators even more popular. The Otis company went on to solve still more challenging technical problems, such as how to make an elevator that would ascend curved tracks up the legs of the Eiffel Tower in Paris. Without the work of Otis, the skyscrapers of the twentieth century would have been unthinkable.

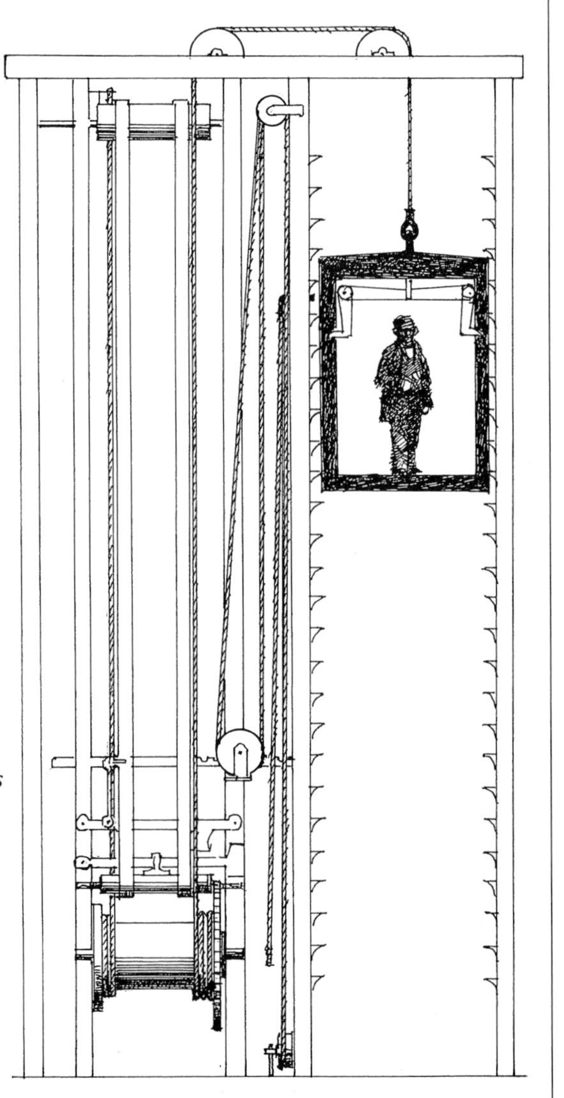

Types of building frame

In the first type of metal-framed buildings, like the Albert Dock in Liverpool (inset), *there was a load-bearing outer wall of brick or stone. Brick was also used in the shallow vaults between the floors. So, although iron was an important structural element, the building kept many features common to earlier structures.*

In structures such as London's Crystal Palace, designed by Joseph Paxton for the Great Exhibition of 1851, metal and glass played a greater and more obvious role.

then, engineers saw how metal construction could help them solve new problems and meet new challenges. Iron and steel could play a major role in creating such 'modern' building types as the factory and the warehouse, the office block and the department store. They also saw how the new technology of building interacted with the other new technologies of the industrial revolution – James Watt's work was a good example.

A different 'modern' building type was the railway station, another representative of industrial revolution technology. Metal construction could help here too. One of the earliest examples was the Central Railway Station at Newcastle upon Tyne, built by John Dobson between 1846 and 1850. Before long, London and many other cities were served by stations along the same lines, and many of these are still being used today.

Technology on show

Like the New York World's Fair, the great European exhibitions of the nineteenth century did more than their fair share to publicize the new building materials. The first of these was London's Great Exhibition at Crystal Palace in 1851. In this case, the exhibition building itself showed some of the potential of metal and glass building. It was designed by Joseph Paxton, who had experimented with this type of building in greenhouses at the Duke of Devonshire's house at Chatsworth, Derbyshire. The Crystal Palace brought metal construction valuable publicity. Unfortunately, it was destroyed by fire in 1936.

Another great exhibition, the one in Paris in 1889, brought metal and glass construction yet again into the public eye. Gustave Eiffel (1832–1923) was the engineer involved. He had already done much influential work in France, but his most famous project would show the people of Europe just how high metal structures could be built. For Eiffel proposed and built a metal tower 984 feet (300 metres) high, which on completion was the tallest structure in the world, and remained so for over forty years. (in 1959 its height was raised to 1,052 feet [321 metres].) The Eiffel Tower caused controversy in Paris – many people did not like its appearance, nor its considerable effect on the city skyline.

Less controversial was the Palais des Machines, the large metal and glass hall, also on the exhibition site. This building was admirable for the way it used vast metal arches to support a roof spanning a wide and uninterrupted floor area. The resulting structure was an ideal exhibition space. Once more it showed how it was possible to build things in metal that were impossible to construct in traditional materials.

The truly enormous trussed arches of the Palais des Machines enclosed a space some 1,375 feet (420 metres) long. Each arch tapered to a point at the bottom, so that only a small area of steel touched the ground. This, and the fact that most of the building was roofed and walled with

Eiffel Tower
This Paris landmark was built between 1887 and 1889.

Forth Rail Bridge
Benjamin Baker's bridge uses the principle of cantilevering to cross the wide span of the Firth of Forth, Scotland. Steel tubes help cope with the huge forces.

Early skyscrapers
In the first skyscrapers of the 1880s and 1890s, such as the Masonic Building in Chicago, built in 1892, the outer brick walls were self-supporting. The steel frame was used to support the floors and roof.

Crystal Palace
Although traditional in its shape, London's Crystal Palace (1850-51) took advantage of the scope for prefabrication offered by the new materials. Pillars, girders, joists, and plate glass were all made to size, in factories as far away as Birmingham, and taken to London by rail.

Empire State Building
For a long time the world's tallest building, the 85-storey Empire State went up rapidly between 1930 and 1932. A heavy steel skeleton is surrounded by concrete and glass. Much of the outer masonry is faced with limestone.

Palais des Machines
This extraordinary structure was built for the Paris Exhibition of 1889. Its roof is supported on vast, trussed steel arches that embrace almost the entire width of the building (one can be seen at the end of the hall in this illustration). Another innovation was the use of a 'skin' of glass fixed to the outside of the frame to make the outside walls.

The new materials
Concrete and steel allow architects and engineers to create structures of new shapes and new sizes. Vast exhibition halls, tall skyscrapers and bridges with longer-than-ever spans are three examples of new building types that supplied the new needs that were felt as a result of the commercial expansion of the late nineteenth and early twentieth centuries. And there are also structures like the Eiffel Tower, still eccentric a century after it was built, which express something of the engineer's joy in the new possibilities these materials were opening up.

glass, gave the structure an extraordinary visual lightness – an illusion, of course, since the tonnage of steel used must have been enormous.

Buildings like the Palais des Machines show the versatility of metal. Used with sensitivity it can provide the basis for buildings that are both good engineering solutions and highly satisfying to the eye. And what was true of these 'industrial' buildings was also true, in another way, of the houses of the late nineteenth-century Art Nouveau movement. Here, the sculptural qualities of metal were brought to the fore, although architects were still capitalizing on the structural benefits to be gained from the use of iron and steel.

Ferro-concrete

While the architects of Art Nouveau were creating their fanciful decorative effects, engineers were investigating another way to use metal in building. This time they were marrying it with a much older building material – concrete. Used by the Romans, rediscovered by John Smeaton in 1774, and incorporated, as we have seen, in iron-framed buildings as a fireproofing measure, concrete had already made its architectural comeback and proved to be an indispensable builing material.

But no one really exploited the structural implications of combining iron and concrete until the end of the century. One of the pioneers of the technique of reinforcing concrete with iron rods was French builder François Hennebique. He it was who worked out the best proportion of concrete to iron, and the optimum arrangement of the rods. He built himself a house out of reinforced concrete, including overhanging rooms and balconies, and other elements that were possible only with such a strong material.

Hennebique's house was an effective piece of publicity, and within a decade reinforced concrete was in common use all over Europe. Architects had discovered a material that gave them the best of both worlds – the flexibility and plasticity of concrete, and the strength of iron or steel. Such qualities were as useful in houses as in large public buildings, and would soon prove invaluable in civil engineering projects, from tunnels to bridges. From the 52-storey Woolworth Building in New York, completed in 1913, to the many more steel-and-concrete buildings that followed it, the new materials transformed our towns and cities.

The story of the adoption of these new materials in the buildings of the nineteenth and twentieth centuries is a typical story of the interactions that make human inventions so interesting. It is a story in which figures from outside the world of building as we normally understand it have an important place – people like Watt and Otis. Many of the other important figures were builders and engineers rather than architects in the usual sense of the word. It is also a story in which transfers of technology around the world provide a key. Without the great nineteenth-century exhibitions, and the great gestures of the period, like the Eiffel Tower, the ideas would have caught on more slowly. And finally, and most significantly, it is a story of new demands producing new solutions. The industrial revolution created a need for new types of buildings, railway stations and warehouses, factories and mills. It soon became clear that technologies that had evolved for other reasons related to the Industrial Revolution could supply these needs too. Such are the interactions, between one country and another, one industry and another, one skill and another, which make the study of human ideas so interesting.

EARLY MEDICINE

Ancient approaches to restoring and maintaining health, little changed over the centuries, laid the foundations of modern medical techniques

Since human life began there has been disease. Sometimes the reasons for the illness are obvious. In the case of an accident resulting in a cut, bruise, burn or fracture, for example, the link between cause and effect was probably apparent even to our earliest ancestors. Other causes of disease, such as poisons, also became clear.

But this left a vast number of diseases of which the cause was far from obvious. Nowadays, science recognizes a host of hidden influences on our health: bacteria, for example, about which early societies could know nothing. So early peoples found other ways to explain diseases caused by bacteria: the sufferer was possessed by an evil spirit, or was being punished because of an offence against the gods, or against society.

Pre-scientific medicine

Early societies evolved various ways of dealing with these mysterious diseases. Many of these early cures seemed to work. This was sometimes because they did no harm, and the body left to itself will usually recover from all but the most serious diseases. In other cases, the success was due to some active ingredient or health-giving practice that did directly help the patient.

Ritual and magic often played a major part in the cures. To modern eyes, such cures seem strange and arbitrary, but the witch doctors of primitive societies were operating within as well-defined a set of conventions as does a Western physician today. When a medicine man did a dance in an elaborate costume, it might be because a spirit needed appeasing; when someone was told to wear an amulet made of an animal's claw, some of the strength of that animal might be expected to rub off on the patient.

Witch doctor

Egyptian physician, c.2600 BC

It is easy to dismiss such practices today, but often they could be helpful to the sufferer. Their usefulness lay first of all in the simple fact that people believed in them. And their belief sometimes gave them the strength they needed to fight the disease. The medicine men were helped in this by the fact that their medicine was often holistic – they were treating the whole person or entire family group.

The effectiveness was perhaps increased because in many early societies it was believed that everything was controlled, or at least influenced by the spirit world. The most pressing factors for human survival – the success of the crops or the hunt, human fertility, and human health generally – were in the control of the gods. Keep the gods happy and you would have a good chance of staying healthy. The village priest and the medicine man were often the same person. In some North American Indian tribes the shamans sat with the chiefs. As well as healing the sick, they captured the souls of enemies, and were feared for the harm they could do if angered.

In any case, it was not all ritual and magic. Over the centuries, primitive medical practitioners built up a considerable fund of knowledge about curing disease. Some early Native Americans even practised surgical techniques, such as trepanning. In particular, physicians of the ancient world discovered that some plants had therapeutic properties, and herbalism is probably one of the oldest successful forms of medicine.

The first civilizations

The cultures of Ancient Egypt and Mesopotamia made some of the most important early contributions to medicine. It was in these civilizations, both flourishing between 3000 and 2000 BC, that medicine started to become more scientific in the modern sense of the word – that is to say, there was a greater emphasis on organized training, on writing down suggested methods and case histories, and on diagnosis based on observation and questioning the patient.

Egyptian doctors were members of the priesthood who had gone through a special medical training – indeed, medical knowledge was often passed down from father to son. They practised a type of medicine that was a strange mixture of primitive and modern. They took the pulse and estimated the body temperature with their hands and, although they did not know what the scientific standard pulse rate or temperature was (or even by what the pulse was actually caused) the information they gathered must have given them some insight into the individual's health. They could use splints for fractures and drain abcesses. Their records give details of surgical procedures, some of which are still followed today. And, like their less scientific predecessors, they had a wide repertory of medicinal herbs that they could use to treat many disorders.

But Egyptian medicine retained some of the characteristics of what had gone before. Still the preserve of the priesthood, it relied heavily on prayers to the gods, and there were several gods that had healing abilities. One was Khonsu, the Moon god, whose statue was supposed to have healed the sister of Nefrure, wife of Pharaoh Ramesses II, who was possessed by a hostile spirit. Another was Imhotep, the great engineer and architect of the Old Kingdom, who was later deified by Egyptian priests. An offering to the priests at the temple of one of these gods was thought to help one's health. Alternatively, one might bypass the priestly system, and pray to one of the gods of Egyptian popular religion. Taweret, the goddess of childbirth, was one of the most important of these. A further remedy, if it was thought that an ancestor's spirit needed appeasing in order to restore health to a sufferer, was to make an offering at the relative's tomb.

Medicine in Mesopotamia was in many ways similar to the Egyptian science. It was also practised by priests, and was a mixture of scientific and traditional methods. Astrology was also reckoned to be an important aspect of healing – the people of Ancient Babylon were among the first astrologers to leave written records.

Acupuncture
According to Chinese thought, the body contains twelve meridians, which can be thought of as channels along which the life force 'chi' should flow. Each meridian is associated with a particular organ of the body. One of the aims of traditional Chinese medicine, therefore, is to influence the flow of chi along the meridians. This can be done by inserting a needle at specific points on the meridians, a practice known as acupuncture. Chinese physicians use an elaborate map of the body that shows the many acupuncture points: each is related to a particular part of the body or to a disorder. The points can be stimulated in various ways: by the temporary or long-term insertion of a needle, by pressure, or by burning a dried herb called moxa on a needle inserted in one of the points. Acupuncture is increasingly popular in the West, where its effectiveness in a range of applications, from pain relief to helping people to stop smoking, has been widely demonstrated – even if Western science still cannot explain how it works.

The Classical world

Ancient Greece, and the Roman Empire that followed it, inherited the medical traditions of Egypt. We know something about Greek medicine from the writings attributed to Hippocrates, a physician of the fourth century BC. In fact, these texts were written by a group of different people, so it is usual to talk about 'the Hippocratic writings'. Much Greek medicine relied upon the theory of the four 'humours'. These were four substances present in the body: blood, phlegm, yellow bile and black bile. Each substance was believed to have a set of attributes, and the four substances were supposed to be kept in balance. A domination of one or other of the humours was thought to cause disease, and physicians aimed to restore the balance.

Hippocrates

If humour theory was spurious, the idea of restoring balance was not. And there were other aspects of Ancient Greek medicine that were effective. Although the main aim of the doctors was to create a balance of the humours in the patient's body, the method used to obtain this was often based on changes in diet. Greek doctors also realized the importance of a good 'bedside manner' and, in particular, of taking a case history.

The diagnostic side of medicine was reinforced in the writings of Aristotle, who stressed the science of medicine. According to Aristotle, treatment should be made on the basis of sound and careful observation of the patient; general knowledge of health and sickness could also be improved by observation – by dissecting, and recording details of anatomy and physiology.

This scientific approach was taken up by the Romans, who inherited and refined many of the ideas of Aristotle and of the Hippocratic writers. We can see how some of these were taken up in the writings of Galen, a highly important physician of the second century AD, who learnt his anatomy while acting as physician to the gladiators' school. Galen's influence was felt for at least 1,500 years after his death (*c.*200 AD).

But perhaps even more relevant is the archaeological evidence about Roman society. The Romans, recent compared to the other civilizations so far mentioned, left a vast amount of material behind them – buildings, works of art, and everyday tools and utensils. From this material, we learn that the Romans had a high standard

Herbal remedies ancient and modern

Many forms of traditional medicine around the world use plants. Now, modern science can often isolate the active chemicals in these plants, and make the chemical synthetically so that it is more widely available. A well-known example is the plant Ephedra. *This is a bush that has been used in Chinese medicine for thousands of years, both to treat low blood pressure, and as a tea made from the twigs to reduce fever and soothe coughs. Scientists have isolated the drug ephedrine from this plant, an alkaloid that is now widely used in the West in many ways – to raise the blood pressure, and to treat hay fever and asthma. Another example is the foxglove* (Digitalis), *which was used by Western herbalists. This has been discovered to contain the substances digoxin and digitoxin, both of which are used in medicines prescribed to treat heart disease. In the case of ephedrine, the drug can be made synthetically, or derived from plants specially grown for the purpose. In the case of digitalis, the chemicals are still taken directly from the dried leaves.*

The four humours

Much early medicine was centred on the concept of the four humours, based on the elements air, fire, water, and earth, and supposed to influence both one's character and one's health. Each humour governed a particular body fluid and a set of characteristics. For example, someone with plenty of air was thought to be a cheerful or sanguine character, whose body contained much blood. If water was the main element, phlegm was thought to be the dominant fluid, and the character was supposed to be sluggish, and so on.

Key

1 *Ephedra (*Ephedra sinica*)*
2 *Foxglove (*Digitalis purpurea*)*
3 *AIR – sanguine, cheerful*
4 *FIRE – choleric, hot-tempered*
5 *WATER – phlegmatic, sluggish*
6 *EARTH – melancholy*

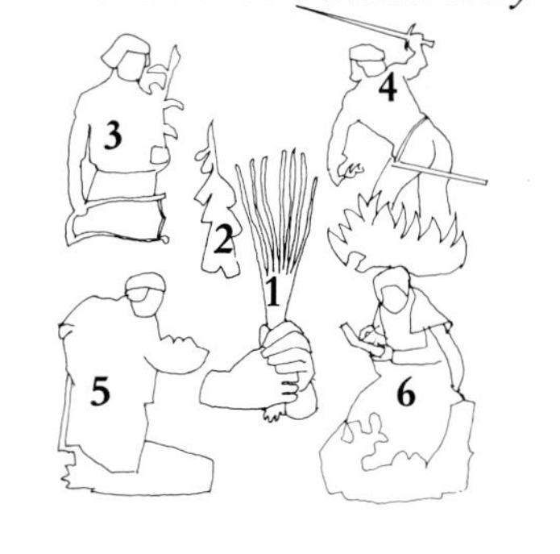

of public health. In their cities at least, they had efficient sewers and clean water supplies. They built hospitals. And they had quite high standards of hygiene. They also developed the practice of surgery, as we can see from the surgical instruments that have survived from the period.

At the end of the Roman period, much of the medical learning that the Greeks and Romans had built up was inherited by the eastern Roman (later the Byzantine) empire, based at Constantinople. And although we know less about the expertise of those who succeeded in western Europe, they must have inherited many Roman medical skills.

Eastern traditions

While the influence of the great Western civilizations continued in Europe and the Middle East, quite different sorts of medicine had evolved in civilizations that existed further east. A good example was China. Here, by the sixth century BC, a quite different system of medicine had evolved. This is a system that has now been used for thousands of years.

One key concept in the Chinese view of life is the idea of the two opposing forces that dominate the world – yin and yang. Everything in the world can be

The Chinese symbol of yin and yang

described in terms of yin and yang. And, just as in Western medicine physicians aimed to attain a balance of the humours, so in China a balance of yin and yang is thought to be necessary for health. Yin is the feminine force, dark and mysterious. Yang is masculine, and is seen as bright and active. Each organ of the body is designated as yin or yang, and internal or external forces are thought to influence the balance of yin and yang in a person.

The Chinese believe that an important way in which an individual's health can be changed is to regulate the flow, through the body, of the life force called 'chi'. Many techniques of Chinese medicine, including acupuncture, are designed to regulate the flow of chi. Chinese doctors also use herbs, massage, and other treatments to attain a balance of yin and yang, and a steady flow of chi. All the treatments are holistic – even though an acupuncture point influences a particular organ, the overall aim is balance throughout the entire body. And prevention is preferable to cure. This is an approach to medicine that is still adopted in China, and is beginning to find favour in the West as well.

Later approaches

But after the end of the Roman Empire, Western physicians knew nothing of China. They continued to practise medicine according to the doctrine of the four humours. The monks of the Middle Ages kept scholarship alive and read what classical writers like Aristotle had to say about medicine. Hospitals in this period were usually monastic foundations, run by monks or nuns.

The medieval world was also open to some influence from other cultures. The Islamic peoples, for example, had some important writers, such as Avicenna, who influenced medicine in both East and West. The ideas of the Classical world were also transmitted through Avicenna, as he often refers to Aristotle and Galen in his writings.

There was also folk medicine, everyday cures that probably went back thousands of years to the period before medicine became a science. One of these was the idea of 'signatures', the principle that like cures like. Thus a plant like lungwort, with vaguely lung-shaped leaves, was used for respiratory disorders. This may seem an arbitrary way to choose a medicine, but behind it were years of trial and error, and if the theory behind the doctrine was flawed, many of the medicines worked. And modern science has picked up on some of the traditional remedies, isolating the chemicals within them that are beneficial.

Indeed, when we look at the medical methods of past cultures there is nearly always something, among the oddities, from which we have learned, and are still using today. Sometimes, as with Chinese medicine, it is an entire system that is still demonstrably working. Sometimes it is a general approach, like the insistence of Aristotle that we should base our medicine on close observation. Sometimes it is a detail, like a chemical in a traditional herbal remedy that scientists now make synthetically, but which is still used to cure people. All these ideas and inspirations are saving lives.

VACCINATION

A revolution in public health was achieved by infecting healthy people with diseases which their natural antibodies fought off

During the eighteenth century in Europe, smallpox was a major cause of death. It spread rapidly and attacked members of all classes. Sometimes there were particularly virulent outbreaks that laid people low on a horrifying scale. And for those who caught the disease but survived, there was the disfigurement of pock-marks to remind them of their suffering for the rest of their lives. The only consolation was that they would be unlikely to get the disease again – the first attack conferred immunity on the sufferer.

This fact had led some physicians to experiment with inoculation, taking infected material from a smallpox sufferer and using it to give someone else a mild dose of the disease, with the hope of protecting them from a fatal attack. But this procedure was risky. The hoped-for mild dose could turn serious, even fatal; moreover, the inoculated patient became a carrier of the disease and could infect others. Smallpox remained a killer.

The person who was destined to change this was Edward Jenner, the son of a Gloucestershire clergyman, who was to become the doctor in his home town of Berkeley. Jenner trained with a local surgeon before going to London as a pupil of one of the foremost physicians of the day, John Hunter of St George's Hospital. Jenner thus encompassed both traditional and modern views of medicine, a combination that was to prove invaluable in his practice and in his work on smallpox.

From his traditional knowledge, Jenner gleaned the local wisdom that, frequently, people who had been infected with the disease cowpox (quite a mild disease that could be caught from cattle) did not get smallpox. What would be the result of inoculation with material from a cowpox sufferer? Would the patient be protected against smallpox as well as the less serious disease?

Jenner knew from his work with Hunter that he had to experiment in order to test his observations and to find out what would happen. In May 1796 he had his chance. A local dairymaid, Sarah Nelmes, was brought to him with cowpox. Using infected matter from this patient, he inoculated an eight-year-old boy, James Phipps, who developed cowpox as a result. Now Jenner had to find out whether the boy would contract smallpox. So he gave the child a further inoculation, this time of smallpox. He did not contract the disease, and Jenner had proved his case. If trying to give the boy smallpox in this way seems a terrible risk today, it must be remembered that inoculation with smallpox was thought to be justifiable in Jenner's time – it was, after all, giving the patient a chance to avoid a fatal disease.

Recognition was slow in coming. Jenner's first paper on the subject was rejected by the Royal Society in 1799, so he published his findings privately. Nevertheless, the worth of the practice of vaccination (the word is derived from the Latin *vacca,* meaning cow) was soon self-evident, and Edward Jenner's methods began to spread around Britain, Europe, and America.

Much work had to be done before other diseases could be eradicated by vaccination. In particular, the work of Frenchman Louis Pasteur, and German bacteriologist Robert Koch (who worked on tuberculosis, cholera, bubonic plague and malaria) put vaccination on a sound footing by the end of the nineteenth century, and transformed public health in much of the Western world as a result. But Jenner started the process, and millions have him to thank for their lives.

Jenner took an enormous risk when he inoculated his first patient with cowpox. It was the sort of experiment that would not have been allowed today; but it was a gamble that paid off, to the ultimate benefit of millions.

1796 Edward Jenner inoculates James Phipps with cowpox (*above*).

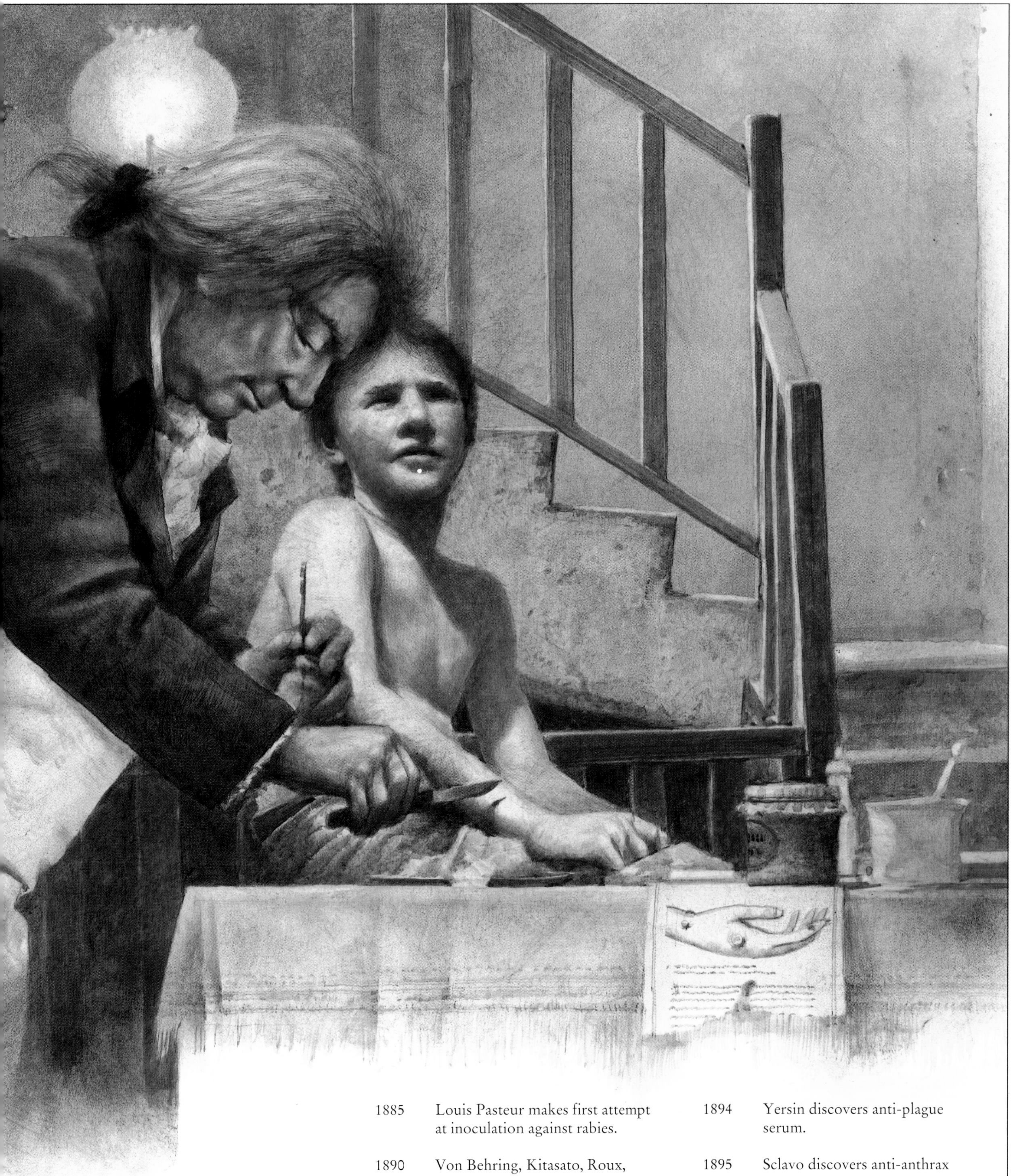

1799 Jenner's paper on his findings rejected by the Royal Society.

Jenner publishes his findings privately; news of his success spreads quickly.

1885 Louis Pasteur makes first attempt at inoculation against rabies.

1890 Von Behring, Kitasato, Roux, Martin, and Chaillou work on preventive sera, which contain antibodies, and can be taken from the blood of infected patients. The first to be discovered combat diphtheria and tetanus.

1894 Yersin discovers anti-plague serum.

1895 Sclavo discovers anti-anthrax serum.

1896 Roux, Metchnikov and Salimbeni discover anti-cholera serum.

ADVANCES IN MEDICINE

Stethoscopes gave doctors new insight to patients, while antiseptics and anaesthetics made surgery safe and painless

The work of Edward Jenner on vaccination went hand in hand with a generally more scientific approach to life. Medicine in general benefited from this approach, with the study of anatomy, for example, becoming more widespread, systematic and accurate. This development went back to the sixteenth century, with the work of the great anatomist, Andreas Vesalius, Professor of Anatomy at Padua. Vesalius built on the work of important predecessors, such as the Graeco-Roman physician Galen, who lived in the second century AD, but whose influence spread through the Middle Ages and beyond. But Vesalius was arguably the first to study scientifically the inside of the human body. Benefiting from the steady supply of healthy corpses available in a time of capital punishment, he dissected away, charting bones, muscles, and inner organs. His lectures were highly popular, and he illustrated them with superb drawings and charts, later published.

What Vesalius did was quite straightforward: he simply took the body apart, looked at it with care, and recorded his findings. In so doing he built up knowledge that proved invaluable to the physicians who came after him, making a map of the body that everyone could follow. At the time this seemed an outrageous thing to do; it seemed to threaten both the doctors of the establishment, whose ancient ideas were being overturned, and the men of the church, who took offence at both Vesalius's methods and his findings. This opposition forced Vesalius to leave Padua and take protection as the Court Physician of the Holy Roman Emperor Charles V. The problem is one that many great men and women of ideas have had to face throughout history: the opposition of powerful vested interests.

Another pioneer who overturned previous notions was the English seventeenth-century scientist William Harvey, a student of the great medical school at Padua where Vesalius had taught, who discovered the circulation of the blood. In retrospect, it is surprising that Vesalius, with his genius for observation, did not understand the circulatory system. But he did not grasp the implications of the blood vessels he dissected, and the breakthrough was left to Harvey. Harvey worked in a way similar to Vesalius, but he dissected a range of different species, and concentrated on the blood system. He studied the action of the heart, noticing how it expelled blood every time it beat, and worked out where this expelled blood actually went. Soon he had discovered the whole circulatory system of arteries and veins.

Harvey's was an epoch-making discovery. It was important not just from the point of view of the blood system, but encouraged a whole new view of the body, in which various substances, from nutrients to oxygen, could be transported through ducts specially designed for them. Harvey's discovery also suggested ways in which diseases could be passed around the body.

The eighteenth century, in which Jenner did his work on vaccination, was also the first great age of the medical schools, with prestigious institutions that would build on the work of men like Vesalius and Harvey being established in centres such as Edinburgh. The number of doctors per head of population was on the increase, and fashions such as 'taking the waters' at popular spa resorts made many people aware of the steps they could take to improve their health.

This trend was continued into the nineteenth century, a period which brought many of the developments that have influenced medicine until today. At the most basic level the coming of modern

In the sixteenth and seventeenth centuries, the study of anatomy became central to medicine, and doctors like Vesalius studied corpses with all the fascination of pioneers.

medicine had to do with a clearer understanding of how the body works, and of what happens when it is diseased. Experiments in laboratories and hospitals all over Europe vastly increased our knowledge of physiology. As scientific instruments, such as the microscope, were improved, so this process of experimentation proved still more fruitful. And as this application of a more analytical approach paid dividends in medical theory, its value also became recognized in practice.

Diagnosis

As physicians began to understand the physiological processes that cause diseases, so they became more interested in understanding patients' symptoms. Previously there had been a very heavy reliance on talking to the patient, whose own perception of the illness was of paramount importance. But in the nineteenth century, a host of tools was invented that made it easier for the doctor to examine the patient and actually look at the symptoms directly. Thermometers, stethoscopes, machines for measuring blood pressure, and devices for looking inside the patient's body were all invented or developed during the nineteenth century. The science of diagnosis grew by leaps and bounds.

One of the simplest diagnostic aids to appear was the stethoscope. Physicians had known of the importance of the rate of the heartbeat since William Harvey published his findings about the circulation of the blood in 1628. It was also discovered that listening to the sounds of a person's inner organs – especially the lungs – could help in diagnosis.

The French physician René Théophile Hyacinth Laënnec used such procedures, but was frustrated when putting his ear to the chest of a rather overweight patient and finding that he could hear very little. Taking a cue from the trumpet-shaped hearing aids that were used widely at the

Stethoscopes from the nineteenth century

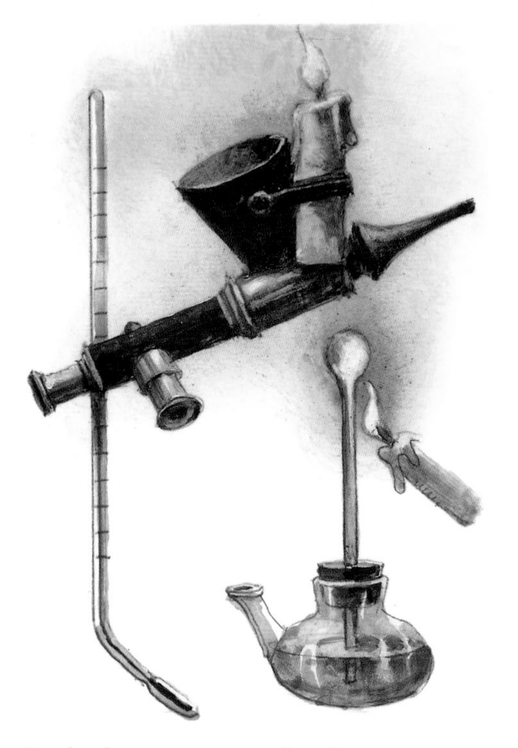

Early thermometer and endoscopes

Killing pain

Before the nineteenth century, people expected surgery to hurt: operations were carried out with the patient fully conscious, or under the influence of a large dose of alcohol and held down by the surgeon's muscular assistants. The beginning of the end of painful operations came when Sir Humphry Davy discovered the gas nitrous oxide in 1799. At first it was a discovery that did not seem very useful. The most notable property of the gas was that it made you laugh when you inhaled it. Showmen made money out of demonstrations of nitrous oxide, making volunteers from the audience roar with laughter. Then, in the early 1840s, one volunteer injured himself in the leg when under the influence of the gas, but felt pain only when the effects of the gas wore off. A dentist in the audience, Dr Horace Wells, realized the usefulness of what he had seen. Later, Wells tried nitrous oxide on himself, inhaling much more of the gas – enough, in fact, to make him unconscious. While he was out cold, a fellow dentist triumphantly extracted one of his molars: Wells felt nothing.

At around the same time a second medical team was experimenting with ether, another substance that people had been inhaling because of its pleasant effects. The physician in question, Crawford Long, lived and worked in Jefferson, Georgia. Like Wells, Long tried the substance on himself, inhaling it from a cloth; he also tried it on his students. Then, one day in

Nineteenth-century physician administering anaesthetic via a face mask.

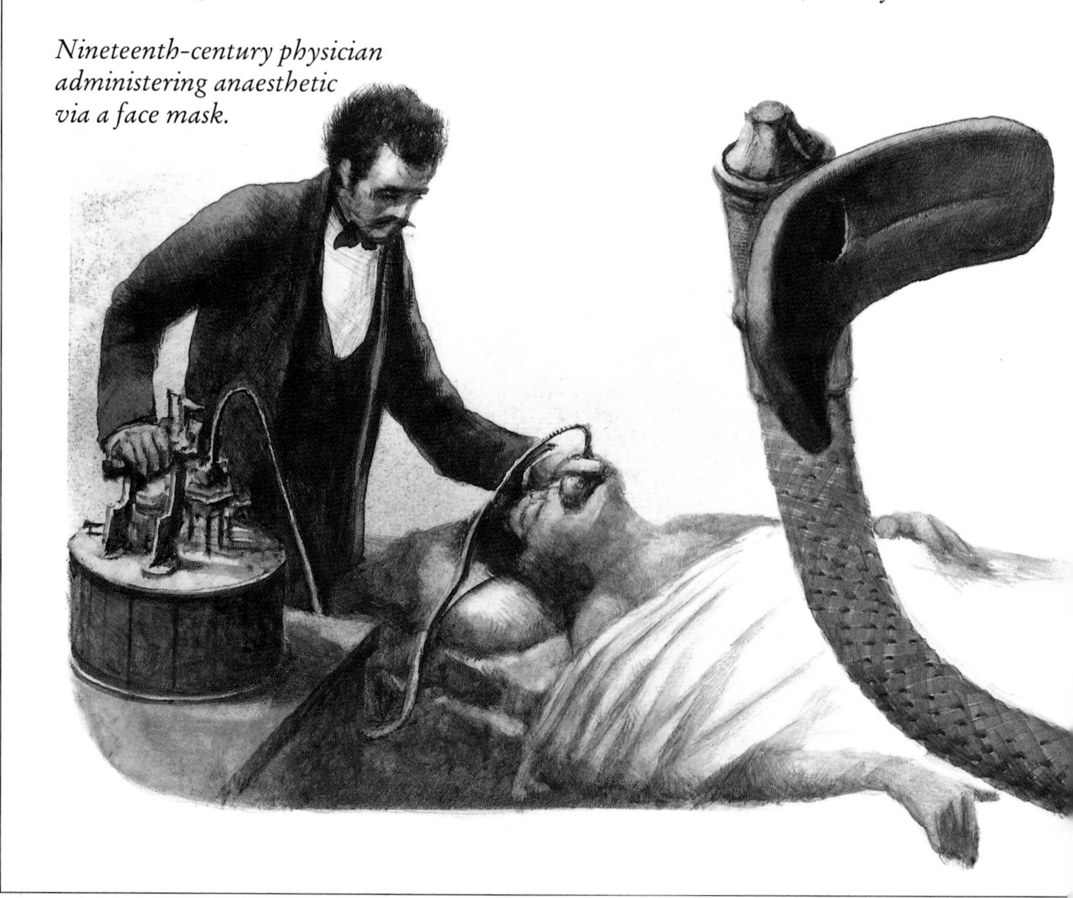

time, he rolled up a convenient piece of parchment and listened to the woman's chest. Laënnec was impressed by the amplification and went on to experiment with other materials. He soon settled on a simple wooden tube, with a circular sound-collector at one end and a wooden earpiece at the other, a device that became popular in the 1830s. Two decades later the familiar modern stethoscope, with its twin earpieces and a flexible tube, had evolved.

The other mainstay of clinical diagnosis, the thermometer, was well developed by this time. (The first true thermometer was developed by Galileo at the end of the sixteenth century.) But the device was not widely used in diagnosis until the end of the nineteenth century, when medical suppliers began to develop thermometers especially for clinical use. These included specially shaped tubes that would fit comfortably under the armpit.

The thermometer was a scientific instrument that gradually became accepted as a vital tool for the physician. A similar transition was made by the devices for recording pulse traces and measuring blood pressure. Particularly important was the sphygmomanometer, the instrument invented by Samuel von Basch for measuring blood pressure.

A final example of the diagnostic advances of this period is the endoscope in its various different forms. These are devices for looking inside the body without surgery. They are essentially an elaboration on the various specula (often no more than mounted mirrors) that surgeons and dentists had long used to peer into the mouth and other openings in the body. The aim was to look deeper, and to see more clearly. To this end, a light source was needed, together with a viewing lens and a device that would pass some way into the patient's body.

Philip Bozzini designed a device in 1804 that was meant to allow one to look

1842, he decided to use it on a patient who had two tumours on his neck. The patient, James Venable, inhaled the ether, and Long removed the tumour. This was probably the first operation carried out under a general anaesthetic.

Another valuable anaesthetic that was discovered in the nineteenth century was chloroform. The doctor responsible for this was James Simpson, a professor at Edinburgh who specialized in obstetrics. Simpson saw particularly the value of anaesthesia in his own field – he was daily faced with the pain of childbirth. Oddly, he met with opposition. There was a school of thought in the nineteenth century that believed there was something desirable about feeling pain when giving birth. Churchmen quoted the Book of Genesis ('In sorrow thou shalt bring forth children'), but countless mothers were thankful for his discovery, and Simpson was finally honoured by the establishment with a peerage.

Mid-nineteenth-century ether inhaler

into the bladder using a candle in a box at one end of a tube, but the light source was not really strong enough. More successful were devices that did not try to get such a deep view into the body. One such was Herman Helmholz's ophthalmoscope of 1851, which gave a good view of the retina at the back of the eye. Other devices appeared at this time for looking into the ear and the throat. Some of these were also quite successful: for example, the first time the vocal chords of a living person were seen was in 1855, and this was thanks to one of these visual aids.

But such tools were much more effective when better light sources came along. This happened in the 1880s, when miniaturized versions of the carbon filament lamp invented by Edison and Swan came into use. These allowed the development of effective endoscopes for looking into the bladder and stomach.

Rapid surgery and first aid found an important place on the battlefield.

Surgery

Before the nineteenth century, surgery was essentially an emergency procedure, carried out at breakneck speed to avoid infection and to minimize pain. Many people preferred to have operations carried out in their own homes to avoid the unsavoury social image of the early hospitals. There was also probably less risk of infection. In the nineteenth century surgery was transformed.

For a start, surgeons began to find ways to reduce the pain. Before the middle of the nineteenth century, the surgeon was expected above all to work fast. The most famous surgeons had themselves timed, and it was not unusual for an amputation to be carried out in between thirty seconds and three minutes. There was no time for any complex work, and all but the simplest operations were impossible.

Then, in the late 1840s, anaesthesia was introduced. Patients could be rendered unconscious by inhaling nitrous oxide (laughing gas), ether or chloroform. Immediately, operations could be performed at a more sedate pace. Surgeons could work more accurately, and blood vessels could be sutured more effectively. Soon surgeons began to try more ambitious operations. Hovever there were still horrendous problems. Many patients died, usually as a result of infection, soon after their operations.

Nevertheless, soon surgery would be revolutionized. The role of micro-organisms in infection would be understood, operating theatres would become germ-free places, and surgery would become much more like the branch of medicine it is now. This development was due to the work of several people.

The first of these was the Frenchman

Louis Pasteur

Louis Pasteur. Pasteur's pioneering work was to do with fermentation. He discovered that this process was due to the presence of micro-organisms. If these were excluded, the process of fermentation would be halted. He soon realized that the same was true for the process of decay and putrefaction.

Another important pioneer was the German bacteriologist Robert Koch. It was Koch who identified and studied the micro-organism *Bacillus anthracis* that causes anthrax in cattle. In addition, he undertook key research on human diseases such as cholera, malaria and bubonic plague, revealing the large extent to which micro-organisms are responsible for human disease.

But it was a British surgeon, Joseph Lister, who saw the importance of Pasteur's work for the operating theatre. In 1865, before Pasteur's germ theory was widely accepted, Lister realized that if Pasteur was right, he had to wage a determined campaign to exclude germs from his operating theatre. The result would then be a marked drop in deaths from infected wounds.

Part of Lister's success was simply this: that he kept the theatre clean, excluding germs in a way that few of his predecessors thought necessary. But Lister added another weapon in his armoury of cleanliness. He was a follower of the work of the Professor of Chemistry at Manchester, F. Grace Calvert, who had been studying the properties of phenol, or carbolic acid. Calvert, working in orthodox scientific fashion on corpses, discovered that applying the acid slowed down the process of decay. So he began to supply the substance to Manchester surgeons for cleaning wounds.

Lister jumped at this idea. He realized that, in conjunction with a clean surgery, carbolic acid could kill any remaining germs around the operation site. So, to put it simply, he developed a three-stage procedure for excluding germs: first, ensure that the theatre is as clean as possible; second, use a carbolic steam spray to create a mist of carbolic acid around the wound during the operation; third, use carbolic acid in both cleaning and dressing the wound.

The result was a radical reduction in deaths from post-operative infection. What was more, surgeons could now operate on areas, such as the chest cavity, previously thought to be beyond the scope of surgery. And finally, operations could be carried out more slowly, since the necessity to close the wound before infection took hold was less urgent. Millions of people had their lives lengthened as a result of these methods. Many others were able to keep limbs that would previously have been amputated.

'Spare parts'

Another area where there was considerable improvement was in the provision of 'spare parts' for the body. Artificial limbs, false teeth, and similar aids for comfort

Joseph Lister

Dentistry

False teeth had been in existence since before the time of Christ, but they were always an uncomfortable compromise. Fittings had to be done from measurements inside the mouth, and these were rarely accurate enough to ensure a good fit. The breakthrough came in the eighteenth century, with the use of plaster moulds to make casts of the inside of the mouth. This provided a much more accurate basis for making the teeth, and this process is still used widely today.

Another problem with false teeth was their cost. The usual materials were ivory for the teeth, and gold for the wires and plates. The eighteenth century saw the introduction of porcelain teeth, and in 1844 Charles Goodyear's vulcanized rubber appeared. This proved to be a plate-making material that was cheap and easy to work, as well as being comfortable to the wearer.

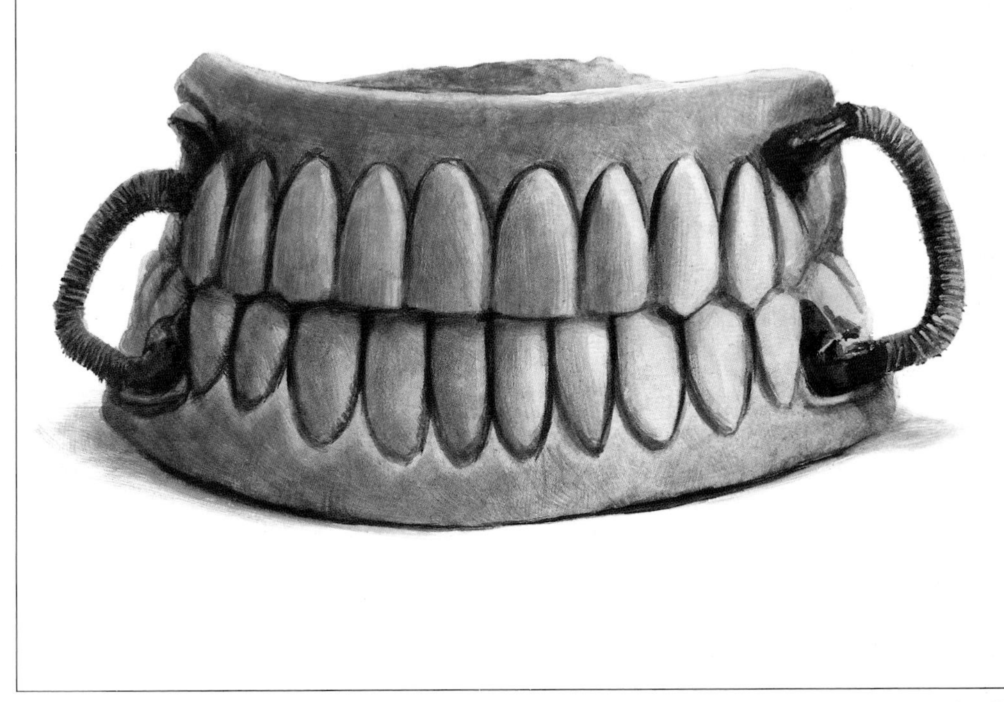

and bodily function were invented or improved during this period. Of course, with amputation such a common operation of the early surgeons, wooden legs had been known and used for centuries – descriptions go back to at least the fifth century BC. But before the nineteenth century, it was rare for artificial limbs to resemble the limbs they replaced, or to be jointed in any way. These spare parts were ugly and unpleasant to wear – it was only the rich who could afford limbs that did anything to counter these shortcomings and even these limbs were hardly naturalistic. Although there had been well-made artificial arms and hands with movable joints, these were usually made by armourers, and looked more at home on the battlefield than in the drawing room.

This changed in 1815 when an artificial leg was produced for the Marquis of Anglesey. This was made to look natural, following the shape of the Marquis's missing limb, and it was jointed at the knee and the ankle. Artificial arms were also improved. A design was made in which a harness attached the artificial limb comfortably, and enabled the wearer to move the arm at the elbow by flexing the shoulder muscles. Designs like these soon became commonplace, and before long realistic limbs like these were available not only to the rich.

Visual aids

Spectacles had also been used for centuries, but came of age in the nineteenth century. Here it was not so much a question of the quality of the finished product, but of getting the correct match of patient and lens. It was traditional to buy your spectacles 'off the peg', often from itinerant salesmen. But in the early nineteenth century it was realized that it was possible to get much better results by testing the individual carefully. So opticians began to use sets of trial lenses to test which gave the best improvement to the patient's sight. Combined with modern diagnostic techniques based on new equipment like the ophthalmoscope, this produced great benefits in the correction of visual defects, setting many of the standards that are still used today.

A lasting legacy

There were many developments in medical science during the nineteenth century, of which those described here are some of the most important. Together they show how a field as complex as the putting right of human disease depended (as it still does) on a large number of people working in different places on different aspects. They all relied on the groundwork done in previous centuries by scientists like Vesalius and Harvey. They often benefited from each others' work – new methods of diagnosis helped physicians who were working on new cures, for example. And they often had to overcome prejudice, whether it was Vesalius failing to convince the churchmen, or the pioneers of anaesthetics meeting with the incredulity of fellow medical workers. Taken together, the work they did represented the most outstanding set of developments in the history of medicine to date. We are still daily in their debt.

Spectacles for close and distant vision

Complementary traditions

Modern medicine, with its highly trained practitioners and high-tech specialisms seems a far cry from the herbal remedies still used in many traditional societies. However, doctors from both worlds are finding that they can learn much from each other – and the prospects for our health improve as a result.

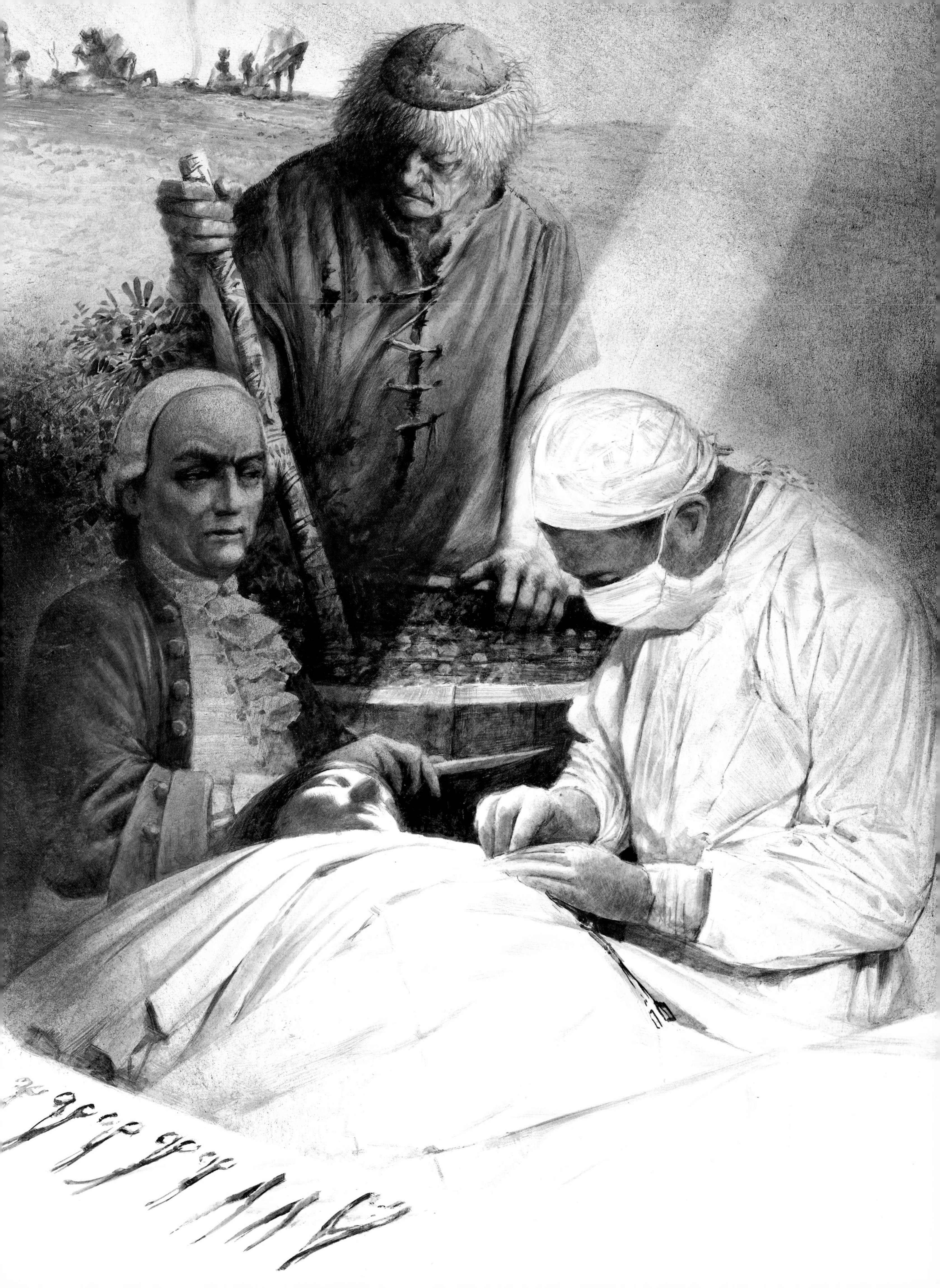

FROM FLINT TOOLS TO PRODUCTION LINES

The human race seems built for success. We have large brains; our upright stance frees our hands for work; the design of those hands makes us dextrous. Yet our hands are not capable of doing everything we want to do. Our bodies seem to need extending to give us greater strength, or still more dexterity. We want to cut up food, to chop wood, to crush grain, to sew, to hunt and to fish. And so people have always used tools.

Other species do too: our close relatives, the apes, show remarkable dexterity with tools, while even some species of birds have been observed using twigs to help them extract food from inaccessible places. But humans uniquely use tools to make tools – a more deliberate, less opportunistic approach to solving the problems of survival.

At its simplest and earliest level, this meant striking one stone with another so that pieces of material were removed and the stone being hit acquired a cutting edge. The first efforts hardly looked like tools at all – simply pebbles with one edge partially chipped away. But tools gradually became more sophisticated and more specialized. Forms such as the disc-shaped scraper and the pointed hand-axe became common. Our ancestors were addressing particular problems, and coming up with working solutions.

During the last glaciation and just afterwards (roughly 75,000 to 10,000 years ago) a variety of different tools appeared. These included the implements for hunting and fishing – fish hooks and harpoons, spears and arrows. Needles and awls, often made of bone, appeared in response to the need to make clothes from animal skins. Knives took on a number of different forms, from long, thin, reaping knives for cutting vegetation, to smaller, finely sharpened flints for cutting meat and skins.

In these ancient artefacts we can recognize the ancestry of many hand tools that are still used. Axes, drills, spindles, chisels, hammers – all are as familiar now as they were in the ancient world. Even a 'modern' tool like a power saw is little more than an ancient implement with an electric motor attached.

But to these tools have been added an array of other devices that have made it easier and quicker to produce things. By the time of the Greeks and Romans, simple machines – wedge, lever, pulley, winch and screw – had been developed. These extended human power, making it possible to move heavy objects with less effort than before.

The modern age has added devices that have actually increased the capacity of production. Among the most influential have been machine tools, items like lathes, mechanical cutters and stamping machines, which perform tasks mechanically, with great speed, that previously took many hours of a craftworker's time. The development of machines has now gone so far, with automation and computer control, that sometimes people seem hardly to be involved in the process at all. Paradoxically, handmade items are now considered more precious than ever: we still value the skill of the craftworker, fashioning an individual artefact, tools in hand.

HAND TOOLS

Primitive wood and metal tools allowed humans to make decorative items, as well as everyday articles

Tools and weapons of the neolithic

1 *Antler axe*
2 *Stone blade in bone haft*
3 *Serrated stone saw*
4 *Bone awl*
5 *Flint arrowheads*
6 *Bone harpoon head*
7 *Flint hand-axe*
8 *Bone barbed fish hook*
9 *Stone adze with wooden handle*

The people of the neolithic created hand tools of great effectiveness and beauty. As described in an earlier section (*see* page 12–17), forms like the hand-axe reached a high degree of sophistication and there was also some specialization – different forms evolved for basic functions, from cutting and pounding, to scraping and piercing.

However, there was a limit to the amount of specialized shaping a piece of flint would take, while still remaining useful. One of the most important early tools for the farmer, for example, was the sickle. To be effective, this was best made in a long, crescent shape with a handle. Some sickles have been found at early agricultural sites, but pieces of flint of the right length were not always available. The Ancient Egyptians had a good solution to this problem. They made wooden sickles instead, in which they inserted a row of small flint blades to make a continuous cutting edge.

Metal and the development of tools

Such tools showed neolithic technology stretched to its limits. Much more adaptability was possible with the use of metal. In most cultures, copper was the first metal to be exploited in any quantity, and used for tools. It was quite a soft metal: this made it easy to work, but meant that copper products were not always strong, and would not maintain an edge.

Metalworkers soon discovered that cold-hammering made the material harder – with enough hammering, copper can be made almost as hard as wrought iron. What is more, some of the earliest copper-workers had another advantage. The early copper mines of the Persian Gulf yielded an impure ore that contained nickel and tin as well as copper. This constituted a natural alloy that could be hammered to produce a still harder material. The equivalent artificial alloy, bronze, and later, iron, provided still better materials for the tool-maker.

Metal tools evolved in various ways. Sometimes they were closely related to what had gone before. The crescent shape of the Egyptian wooden sickle, for example, reappeared in copper and bronze all around the Middle East and further afield. However, the development of tools through these early periods, and then into the Middle Ages is perhaps best illustrated by the various devices that evolved for working wood.

Tools for boring

Neolithic tool-makers had made sharp points of bone, antler and stone for making holes. Metal allowed narrower, though not necessarily sharper points, and a metal awl fitted to a strong wooden handle made a tool that allowed one to apply considerable pressure and make an effective hole in many materials. Another way of inserting the point is with a bow drill, which is rather like the devices early peoples used to make a spark to start a fire.

But a simple, pointed tool only pushed its way into the work – a better solution was to create a tool that actually removed some of the wood. So two-edged bits and spoon bits evolved, both of which removed material as they were inserted. Such bits have been found among Ancient Assyrian tools and were still being used by the Romans.

Bow drill and selection of bits

Woodworking tools

In the ancient world the skills of the carpenter were even more in demand than they are today. A vast range of objects, from houses and ships, to smaller items like wheels and buckets, were made of wood, or had considerable wood in their construction. Even machinery, from the lathes of the Egyptians to the water-mills of the Middle Ages, were made of wood. The products of the carpenters were everywhere, and woodworking tools were highly influential.

Axes and adzes were, for centuries, the most important tools for cutting and shaping wood. They were made in metal in similar shapes to the stone ones that had gone before. But a metal axe was often easier to attach to a handle than a stone one. A beaten-metal axe- or adze-head could be made with a pointed tang at one end, which could be lashed to the handle with hide thongs, or simply driven into it, a design still used in files today. After the introduction of casting, a still better way of joining head and haft was devised – a hollow socket was incorporated into the cast axe-head, allowing one to insert the handle quite securely.

We are not sure exactly when these forms evolved. In the often damp conditions in which archaeological remains are buried, iron, the more 'recent' metal, will decay more quickly than bronze, so we have lost many early iron tools. But an important find of a group of Assyrian iron tools of the eighth century BC shows that many of the classic forms had evolved by this date. Among the tanged tools are a rasp with an irregular grating surface, and a saw blade with teeth sloping back to cut on the pulling stroke. There are also chisels with tangs and metal ferrules for fitting on to wooden handles. And there are two bits that may have been used in wooden handles, or possibly in a wooden brace.

Roman tools

Our most detailed knowledge of tools in the ancient world comes from the Roman period. We have less evidence from Ancient Greece, but it is likely that with tools – as in so many other areas – the Romans borrowed much from their Greek predecessors. Some of the Roman forms are similar to those in the Assyrian find, but many developed significantly in the intervening centuries. There were better saws, for example, with teeth that were set so that the sawdust could be removed continuously from the cut. Roman frescoes depict frame-saws and bow-saws, but these have no way of adjusting the tension on the blade, a facility that would appear in the Middle Ages. In addition to the bow drill, which was known to the Ancient Egyptians, the pump drill was used. The draw-knife appeared for the first time in the Roman period, as did the plane. And, as in Assyria, there was a good range of axes, hammers, adzes, chisels, rasps and bits (again, it is not known whether the latter were used in braces). Indeed, the range of tools in the Roman carpenter's shop was very similar to the hand tools used by a carpenter today.

The diversity of these tools is reflected in the variety of work done by Roman craft workers. There seems to have been much more specialization than previously, with woodworkers concentrating on particular products – carriage-makers, casket-makers, chest-makers and couch-makers are all mentioned in the literature of the time.

Later developments

But there are some notable omissions in our knowledge of Roman tools. There is no evidence that the Romans used the vice (or vise). They used screw-presses for wine, and may have made similar devices to hold work in position on a bench, but we have no proof of this. The vice, as we know it, seems to have appeared in the late Middle Ages, for wood- and metalwork alike. The medieval period also gives us the first hard-and-fast evidence of the brace as a device for turning drill-bits. Another similar device was the T-shaped augur, which used the cross-bars of the T-shape to give extra leverage when boring wide-diameter holes. A further development in the Middle Ages was the increase in the sheer specialization of tools. Saws, axes, adzes, chisels, and rasps appeared in a host of variations for different jobs.

The advances of the industrial age were more in the direction of mechanization and mass production than in the design of hand tools. But there were occasional exceptions. More accurate castings and other metalworking techniques made items like the adjustable metal plane possible. This tool was created by Leonard Bailey of Boston, Massachusetts, in the mid-nineteenth century. It was impressive, not only for its perfectly flat base and its cutter of uniform thickness, but also for its ease of adjustment. There was a nut to move the cutter – on earlier wooden planes you had to use a hammer.

If the power tools of today seem remote from the carpenter's tools of the ancient world, the hand saws, chisels and hammers with which they share the bench go back in a direct line to those ancient forebears. They are testimony to the experience and expertise of generations of ancient craftworkers.

The medieval workshop

Most of today's basic hand tools – hammers and chisels, saws and planes, braces and bits – were available to the carpenter of the Middle Ages. Some were based on Classical predecessors; others, such as the vice and probably the brace, were inventions of the Middle Ages. The wide range of implements was appropriate, since in this period nearly everything was made of wood.

Lathe powered by a bow – a horizontal variation of the bow drill.

Rasp

Plane

Bow drill

Axe

SIMPLE MACHINES

Labour-saving devices dating from pre-biblical times have needed little improvement since first invented

The Ancient Greeks distinguished five basic types of machine for lifting: the wedge, the lever, the block with pulleys, the winch or windlass, and the screw. These have very early origins, and were all mentioned in the first century AD by the Ancient Greek writer on mechanics, Hero of Alexandria. All five devices reduce the physical effort required to do a particular task (although, of course, the actual amount of energy needed stays the same). All of these simple machines attracted much attention in the ancient world. Stories began to circulate about what could be done with them. Archimedes is said to have launched a ship single-handed by using a combination of levers and pulleys. The wedge and lever are probably the earliest of all machines.

The lever

We use levers every day. Each time we open a door, drive a car, use a crowbar, or cut something with a pair of scissors we are using a device that contains a lever. And each lever is making the job easier for us, allowing us to perform tasks that would require much more force, or might even be impossible without the lever.

We know from the monuments they left behind that prehistoric peoples were able to move very heavy loads. It is easy to imagine them discovering the power of the lever very early on. Wherever timber was available, wooden levers would be pressed into service to help shift heavy blocks of building stone.

But it was not until the civilization of Ancient Greece that scientists calculated how levers actually work. It was the Greek scientist Archimedes, in the third century BC, who showed that only one quarter of the strength is required to lift an object if a lever with arms in the proportion 1:4 is used. Archimedes also demonstrated the trade-off – that one has to move the long arm of the lever much further than the object itself will move on the

Using a wooden lever

short arm; using the same example, if the object is to move one foot (thirty centimetres), the operator must move the lever four feet (1.2 metres).

The Greeks and Romans were well aware of the power of the lever. One example of their widespread use of the device is in the ranks of oars with which their ships were propelled. They also used levers in machinery, for example, in the beam presses used to extract the juice from grapes for wine-making.

There are various types of lever, with the fulcrum (the pivot point) placed in different positions. The see-saw design, with the fulcrum in the middle, as in the oar, developed early, as did the arrangement used in the wheelbarrow, with the fulcrum at one end and the load in the middle. Wheelbarrow-like devices are depicted in Ancient Sumerian art of c.3200 BC, showing the antiquity of this idea.

Foot-operated levers, often known as treadles, probably originated in the Far East. These appear on many machines today, and have long been used in hammers for pounding rice, and as part of the control mechanism for weaving looms.

The wedge

Another source of leverage, the wedge, also proved its worth on early building sites for moving and lifting stones. Pushing in a wedge-shaped block of wood or stone was often simpler than applying a lever. And pushing a building stone up a sloping wedge, perhaps with the help of wooden rollers, was an effective way of raising it.

The Egyptians had an efficient way of using wedge-shaped ramps to raise obelisks. In fact, the obelisk was lowered from the top of the ramp into position at the bottom. When the base of the obelisk had engaged in a groove in the plinth at the foot of the slope, the obelisk, now at forty-five degrees to the vertical and delicately balanced, could be pulled upright easily, with ropes. With their colossal style of architecture and their lack of technology apart from ramps, ropes, and levers, the Egyptians often used wedge-shaped ramps to move blocks of stone. However, we do not know the exact form these ramps took.

The wedge was also invaluable for cutting. The Egyptians pushed wooden wedges into holes drilled in slabs of stone. Water was then poured on to the wood, and as the wedge expanded, the stone split. The Greeks used cork wedges in a similar way to split blocks of marble. Woodworkers still unthinkingly use wedges for this type of activity today – the axe and the chisel are types of wedge used to split or remove wood.

The screw

It is possible to think of a screw as a continuous inclined ramp or wedge. It is one of today's most familiar fixing devices. We find it holding pieces of wood together in furniture, or in the assembly of machines and devices everywhere. We take it for granted. But the use of a threaded screw for fixing evolved comparatively late. The spiral of the screw was used in other ways in the ancient world.

The Greek scientist Archimedes (*c.*287-212 BC) is famous for his description of the screw, which was used in his time for raising water. The liquid would be forced up the spiral of the screw (which was contained in a wooden tube) as the operator turned a handle. It was an easier and more continuous action than carrying the water up in buckets.

The screw was in use in the Middle East long before Archimedes' time. It probably originated in Ancient Egypt, where the need for good irrigation in a terrain dominated by alternate drought and Nile flooding was paramount. The Romans also used screws to raise water – it was their solution to a problem that would remain until the age of the steam engine: how to get water out of mines.

Screws were also useful for machines that included some movable part that had to be adjusted, tightened, and secured in place. The Greeks exploited screws on winepresses, for example, and this technique was also used on the oil presses at Pompeii. However, this use did not come into its own until the engineers of the Industrial Revolution found a way of cutting screw threads accurately. The most familiar modern application of the screw – as a device for holding together structures from electrical plugs to furniture – is also a comparatively recent one. The wood screw appeared in the eighteenth century as the result of a gradual process of evolution from the nail.

The pulley

The Ancient Egyptians grasped the principle of the pulley, but did not actually use anything we would recognize as a block and tackle. Much of the work of hauling the large blocks of building stone they employed was done using ropes. They seem to have discovered that they could reduce the amount of work required by passing the ropes around a vertical post or over a post with a curved top. The ropes would be dampened to make them run more smoothly, and the workers would haul away. It would still be backbreaking work, but it was somewhat easier than pulling a straight rope.

Round and round, up and down

One of the greatest challenges to the machine-maker of the ancient world was that of transferring rotary movement into linear motion. The solution lay in that most simple of simple machines, the inclined plane. It is possible to think of a screw as an inclined plane arranged in a continuous spiral formation, and it was this arrangement of the inclined plane that solved this problem. Its most famous ancient incarnation was Archimedes' screw, which allowed you to turn a handle to raise water vertically up the shaft. When the principle was applied to gears, the result was a worm-gear meshing with a circular cog, with the shafts of the two moving at right-angles to each other. Later, this concept was applied to the screw as a fixing device.

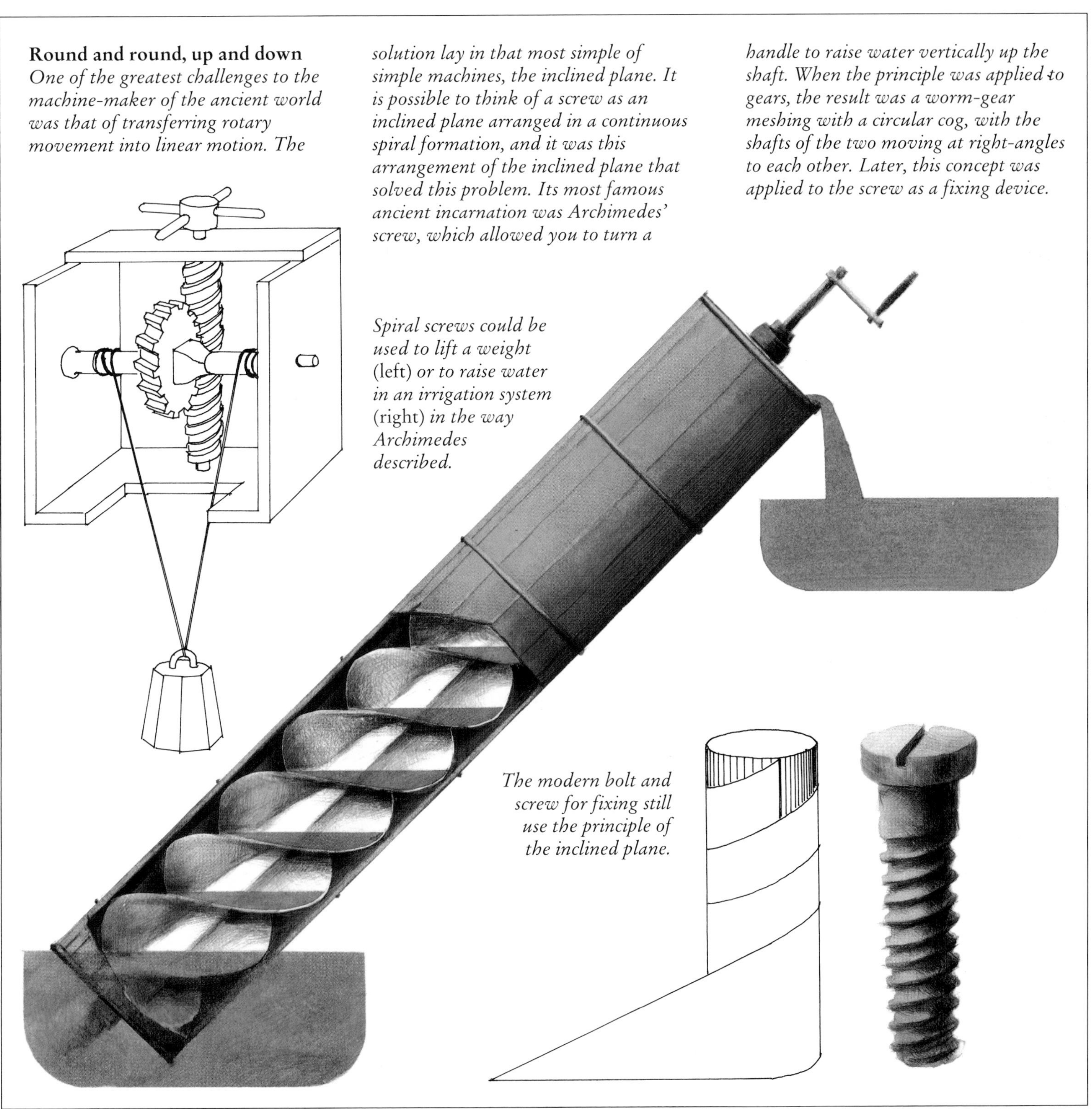

Spiral screws could be used to lift a weight (left) *or to raise water in an irrigation system* (right) *in the way Archimedes described.*

The modern bolt and screw for fixing still use the principle of the inclined plane.

The task of pulling ore up to the surface from the bottom of a mine was made easier with pulleys, notably at the famous Ancient Greek silver mine at Laurion, which was worked from the sixth to the first centuries BC. The key was not only to use pulleys rather than the Egyptian system, but also to use several pulleys in combination. The more pulleys you used, the

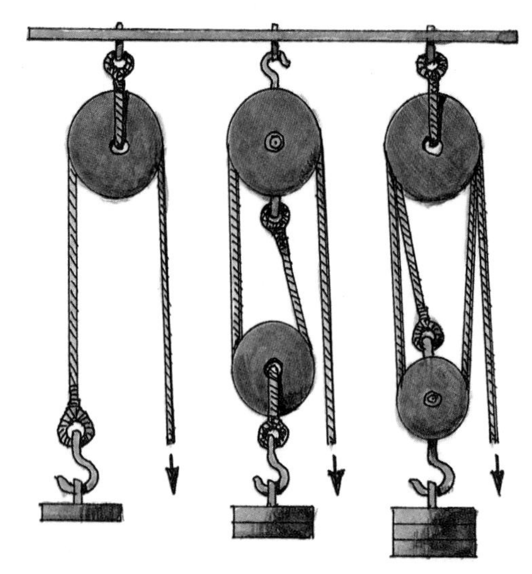

Adding more pulleys saves effort.

more effort you could save. But as with many of the first machines, early construction methods meant that these pulleys were not effective as they might have been. This was because excessive friction was created, forfeiting much of the saving of physical effort.

One other elaboration of the pulley was to build it into a crane. Attempts to do this have been found at least as far back as the Romans and, during the Middle Ages, the Roman design, with power supplied by a treadmill, and a tall gib with numerous pulleys, was still used. There was a famous crane used for unloading boats at Bruges. There is no doubt that the tremendous flourishing of towns and trade during the Middle Ages meant that other cranes were built at those centres where large quantities of heavy goods needed to be taken on and off ships.

The winch

Combining the roller with the lever produced the winch, another machine that was invaluable in ports and on board ship during the Middle Ages, but which originated long before. Again, we know from the writings of Hero that the Ancient Greeks used winches. Indeed, it is extremely unlikely that they would have been able to build their temples without them, since, in contrast to the Egyptians, the Ancient Greeks did not have vast work-forces on which to draw.

No one knows for certain when the winch was first used. An example has been found in a Bronze Age copper mine in Austria which is thought to date to *c.*600 BC. But the device probably goes back long before that. Since then it has been used everywhere that heavy loads need to be hauled: on building sites, on board ship (for raising the anchor or controlling the rigging), for any kind of moving of heavy loads. The addition of the crank (a specialized form of lever) made the winch still more efficient and easy to use. The crank seems to have appeared quite late, probably in the Middle Ages.

But, as in many things, the benefits did not accrue to all people equally. Machines needed skill to build, and so they put the artisan in demand. But often they did not need skill to operate, so the ordinary labourer perhaps benefited only in that his muscles were less sore than before. Machines also put more power in the hands of individuals. For example, a merchant who owned a crane could have bigger cargoes off-loaded more rapidly from his ships. He would be richer and more powerful in consequence.

So, the history of the early machines is, like many of the stories in the history of technology, a tale of double-edged advantages. It is one that people have seen repeated in the more complex machines that have come later, and are still being introduced.

People and machines

Nowadays we are quick to perceive the labour-saving qualities of these simple machines. In the ancient world, where labour was far cheaper than it is today, it was not simply a question of saving labour costs or adding to the comfort of the workers. These machines gave ancient engineers and architects the chance to do things that would have been otherwise impossible – to create bigger buildings, to lift heavier loads, to clear sites that would have remained undeveloped, or to shift cargoes that would have stayed at their place of origin. In this sense, such basic machines as these truly contributed to the development of humanity and the scope of human achievement.

We are not used to thinking of the later Middle Ages as a period of high technology, but a glance at some of the buildings the period has left behind is enough to convince one of the technical sophistication of the times. Cranes, pulleys, toothed gears, levers and winches were all in use, although most of the machines themselves were made of wood and have long since disappeared. The usefulness of such devices stretched from woodworking to shipping, and architecture to trade.

MASS PRODUCTION

A philosophical concept that revolutionized manufacturing and brought 'luxury' items within reach of ordinary working people

The advances – in both the production and use of tools, and the application of steam power to manufacturing – made during the Industrial Revolution were enormous and influential. They were in part, at least, a response to an increase in population and thus also in demand, in part the result of the ambitions of capitalists to make money. Both of these powerful forces would be further fuelled by another development: mass production.

Mass production is something the developed world takes for granted today. But in the eighteenth century only a few people could grasp the concept. In the simplest terms it involves two essential developments. First, many of the manufacturing processes normally carried out by hand are mechanized, which speeds up the production process, allowing greater capacity. Second, the individual parts of a product are made to be identical, so that they can be interchangeable; this leads to speeded-up assembly and much easier maintenance. Both of these developments usually have another consequence, which is that the individual processes involved in making an item are divided up and carried out by specialist and less-skilled workers – or specialist machines. Instead of one artisan making every part of, say, a clock, and then assembling the whole thing, one person will make the gears, another the springs, a third the face, and a fourth the case, while a fifth might assemble the finished item.

From this description it can be seen that the new form of production involved a radical change in the way work was done: it was very different, but very logical. And so it is not surprising that some of the first people to imagine such a system were influenced by the logical thinking of the Enlightenment in France. One of these people was a French General, Jean-Baptiste de Gribeauval, who, in the 1760s, advocated the production of small arms with interchangeable parts. Another was

Thomas Jefferson

Honoré Blanc, who actually attempted to produce guns in this way with the support of Gribeauval. Because of the French Revolution and the death of Gribeauval, Blanc met with little success, but not before he had imparted his ideas to one of the most influential men of his time, Thomas Jefferson, who became President of the USA in 1801. And it was Jefferson who was one of the main people responsible for airing these ideas in the very nation where mass production would find its first, and for a long time greatest, home.

The American gun-makers

At the beginning of the nineteenth century armourers and soldiers in the USA were keen on the idea of interchangeable parts, seeing the advantages of both ease of repair and speed of assembly. Before long, contractors were working for the US War Department to produce arms in this way. Two of the first to succeed were Eli Whitney and Simeon North. They both became well known, but for different reasons.

Eli Whitney had already made a major contribution to the history of manufacturing before he entered the field of mass production. While staying in Georgia in 1792, he was told that cotton planters dreamed of a machine that would remove the seeds from cotton fibres. He duly set to work, and by the following spring had made a machine that would do just that job – the cotton gin. It was simple and successful, and Whitney patented it. It transformed the cotton industry, allowing cotton to be cleaned fifty times faster than before, making viable strains of cotton that had previously not been planted in the cotton fields of the American South, and spearheading a vast expansion in American cotton growing.

But Whitney was a victim of this success. Others pirated his invention and he was left in financial trouble. So he came up with a new proposal that he hoped would solve his problems: he proposed to make 10,000 guns for the US Government, using a mass-production system and unskilled labour. It was a daring scheme, but Whitney underestimated the amount of skilled labour he needed and he delivered late. By the time he had signed his second contract with the government, other arms manufacturers were beginning to adopt the techniques of mass production.

Whitney was a good propagandist for the new system of production, but underestimated the problems. He was not a successful manufacturer, but he spread the word. Simeon North was more careful; he worked out exactly the high standards of accuracy that were needed to achieve interchangeability, and he came up with the goods. He also developed the milling machine, which would later be used in many other industries.

Eli Whitney astride a Colt revolver
The manufacture of revolvers, particularly by Samuel Colt, was one of the most influential early examples of mass production with interchangeable parts.

If North's work showed the importance of precision in production, this was confirmed in another of the influential American armouries, the one at Springfield, Massachusetts. Here they developed a system of gauges for measuring the parts at every stage in the production process. Everyone in the factory had a set of gauges to check their work, and master gauges were kept so that the working gauges themselves could be checked periodically, lest they should become inaccurate through wear and tear.

The Springfield gauges were just one example of how the need for precision was recognized and dealt with at all times. Another was the problem of fixtures. Every time an object is fixed in a tool to be worked, there is a possibility that inaccuracy will creep in. The best way to avoid this is to fix each part in exactly the same way at exactly the same place. To do this accurate fixtures – devices that hold the workpiece securely to the tool – need to be designed. It was John Hall of the Harper's Ferry armoury in West Virginia who first understood the importance of designing fixtures so that inaccuracies would be less likely to creep in.

Hall was probably the first who could claim with justice that he was able to manufacture – in an economical way – guns that were exactly alike, with workers who required little in the way of extra training. Many others would follow, of whom the most famous was Samuel Colt. At his factory at Hartford, Connecticut, Colt further developed the manufacturing system of Hall. He was fortunate in his staff, among whom was the inventor William Ball. Ball created numerous machine tools (including milling machines, screw machines, and presses) that allowed Colt to produce arms faster and more efficiently than before.

Because Colt gave his name to a famous and enduring range of firearms, his name has been closely associated with mass production. Yet his role can more rightly be seen as the manufacturer who took up and capitalized on the methods of his predecessors. And he did encourage talent. It was not only Ball who found employment with Colt. Numerous men who would become prominent American manufacturers, including both Francis Pratt and Amos Whitney, were employed by Colt. The gun, in an odd way a symbol of the independence of the American, gave many American industrialists their own first taste of independence.

Other areas of mass production

Samuel Colt, with his pioneer employees, thought that anything could be made mechanically. In the mid-nineteenth century a host of American manufacturers started to try to prove him right. Nail-makers, barrelmakers, pinmakers, producers of knives and swords, manufacturers of clocks and locks, all of these and more began to turn to mechanization at this time.

One of the secrets was to find a ready market. The gunmakers were successful because they had the backing either of the army, or of the many American individuals who wanted their own arms. The mass-produced McCormick reaper was sold in large numbers to a veritable army of North American farmers. Another successful industry was clockmaking. Clock manufacturers were able to capitalize on a network of pedlars who could sell the products directly to the people. Mass-produced clocks were sold widely all over the USA through this network. And this was the case even though the mass production was not as accurate as that used for firearms, nor the assembly, so far as we can judge, as efficient.

Other industries where mass production took hold were the manufacture of sewing machines and bicycles. Both were inventions of the nineteenth century that had the power to change people's lives. Both appeared first in Europe, but saw mass production in America.

Of the sewing-machine manufacturers, not all, even the famous Singer company, employed mass-production techniques from the start. But those who did met with the greatest success. A case in point is the firm of Wheeler and Wilson, which began in the 1850s using traditional methods. In 1855 they made 1,171 machines. By 1859 they managed an output of 21,306. This increase was due to moving to a more modern plant, and employing a manager who was an expert in mass production and who had learnt his business with Samuel Colt. He managed to produce all his parts mechanically except the needles. The company's ambitious target of 100,000 machines per year was achieved in 1871 and for several years afterwards. Other sewing-machine producers, such as Singer, followed in Wheeler and Wilson's footsteps to mass production, often as a result of the increase in demand they brought about.

The mass producers forced many of the smaller sewing machine manufacturers out of the business. Some of these turned to a new area in the 1880s – bicycle production. There was something of a craze for bicycling in the 1880s and 1890s, and most parts of a bicycle lent themselves to mass production. The exception was the wheels, in which each spoke had to be inserted and adjusted by hand. One wheel

Stock-making

One of the keys to mass production was creating the right machine for the job. A task that used skilled labour was that of carving wooden gunstocks – difficult to mechanize because of the irregular shapes. Thomas Blanchard solved the problem for the Springfield Armoury in 1822 with his profile-tracing lathe, one of several machines that shaped stocks out of pieces of timber. Such a lathe could trace any irregular shape, and later, items from shoe lasts to axe handles were made in the same way.

Singer sewing machines
The sewing machine was one of the great wonders of the nineteenth century. Suddenly, a task that had taken a skilled seamstress hours of skilled, finger-punishing work, could be carried out in a few minutes. Not surprisingly, there was an enormous demand for the new machines. In the end, all the manufacturers realized the benefits of mass-producing their sewing machines.

The most famous of these companies was Singer. They did not adopt mass-production methods immediately, but when they did, it was in a large modern factory at Elizabethport, which was the envy of their rivals. Some of the most important areas of the Elizabethport factory are shown below.

Top row, left to right: *First there was a large foundry, where the sewing-machine bodies and other large parts were cast. Rows of identical moulds stood ready for the hot molten metal.*

Next was the other heavy metal working area, the forging shop. Here, large steam hammers helped the workers to beat or stamp metal parts into shape.

Banks of lathes filled the screw-cutting department at the Singer factory.

Bottom row, left to right: *There was a whole area devoted to the manufacture of needles – parts that would often need replacing on the sewing machines.*

Another area housed various machinery to do with finishing the various metal parts. The equipment included rows of motorized polishers.

Finally, there was the assembly and inspection area, where the machines were put together and tested doing a variety of jobs.

Rover Safety bicycle, forerunner of the modern bicycle

Henry Ford

Henry Ford began his working life as a farmer, became an engineer, and ended as one of the most successful manufacturers the world has known. His career therefore made him aware of the need for automobiles and the most efficient ways to make them. He was influenced by the bicycle industry – one of the first to adopt mass-production – and his first vehicle, with its spoked bicycle wheels, was called the Quadricycle. Twelve years later Ford came up with the car that would change America, the Model T. It was basic, but cheaper than most cars of the time, and its interchangeable parts made maintenance straightforward. Above all, mass production made it available in numbers that could supply the demand. As the benefits of mass production began to be felt at the factory, the unit cost went down. A Model T cost $850 in 1908, but only $360 by 1916. By 1927, some 15 million vehicles had been made. Ford was aware of the effects of mass production on his labour force. The assembly line could be a soul-destroying place, and workers were expected to increase productivity on a continual basis. So Ford introduced the first profit-share scheme, to return to his workers some of their lost incentive.

could take an hour to put together – during which time a worker in a contemporary arms factory could put together some twenty muskets.

So the bicycle represented technology at the turning point. Most of its parts could be mass produced, but assembly was still a stumbling block. The bicycle highlighted the fact that the logic that had been applied to producing parts had not yet been applied to assembly. This would not happen until the coming of a craze more widespread and more influential than that for the bicycle.

The coming of the automobile

The USA is a large, sparsely populated country. The railroad provided long-distance transport; the bicycle helped in the cities. But there was still a great need for individual transport for hundreds of thousands of small farmers and other country people in the Midwest. The invention of the motor car in Europe seemed to offer a solution to the problem, but at first, cars were handmade luxury items. Find a way of mass producing automobiles and you would make a fortune in America.

The problems were similar to those of the bicycle. Parts could be made with relative ease, but assembly was a complex, labour-intensive process. One of the pioneers was Russell Olds. By the beginnng of the twentieth century he was making Oldsmobiles by contracting specialist firms to make car parts. These were then delivered to Olds's assembly plant and trundled round on trolleys to the assembly workers.

But the most important pioneer of mass production was Henry Ford. He turned Olds's production methods upside-down by creating the first true assembly line. Workers were stationed at various points on the line, and the embryo automobiles were moved from worker to worker, gradually acquiring more parts and looking more like the finished product at each stage. Before the introduction of the assembly line it took Ford's operatives twelve hours to build a single vehicle. With the assembly line in position they could do the job in an hour and a half.

The results are well known. Ford created the Model T, the first truly mass-produced automobile, which sold in its millions. In fact, the Model T became the most successful car of all time. By 1925, half the automobiles in the world were Model T Fords. Some fifteen million Model Ts had been produced by the time the company decided to abandon the vehicle in 1927.

The Model T succeeded because it

filled a gap in the market. Vast tracts of North America were populated by farmers who needed basic, reliable transport to get them to town and to take supplies across their land. These were not rich people. Before Ford's revolution they could not afford a car of their own. But by using mass production to bring down the unit cost, Ford opened up this market.

What was more, the farmers liked what they bought. The vehicle that became affectionately known as the 'Tin Lizzie' was immensely practical and reliable, but when it did go wrong interchangeable parts meant spares were straightforward.

Ford gave America the individual transport it needed – and made his fortune. And where Ford had been, others followed. The Model T helped to spread interest in the automobile and other firms began to use mass-production methods. One consequence of this was that companies tended to get larger – mass production worked best in large factories, where there was enough scope to employ specialists to work on the different processes. And so many small car manufacturers, such as Buick, Cadillac and the pioneering Oldsmobile, were absorbed into General Motors, which eventually grew larger than Ford itself. The assembly-line method of production was eventually adopted in countless other industries all over the world. Mass production became the rule rather than the exception. It was a manufacturing revolution.

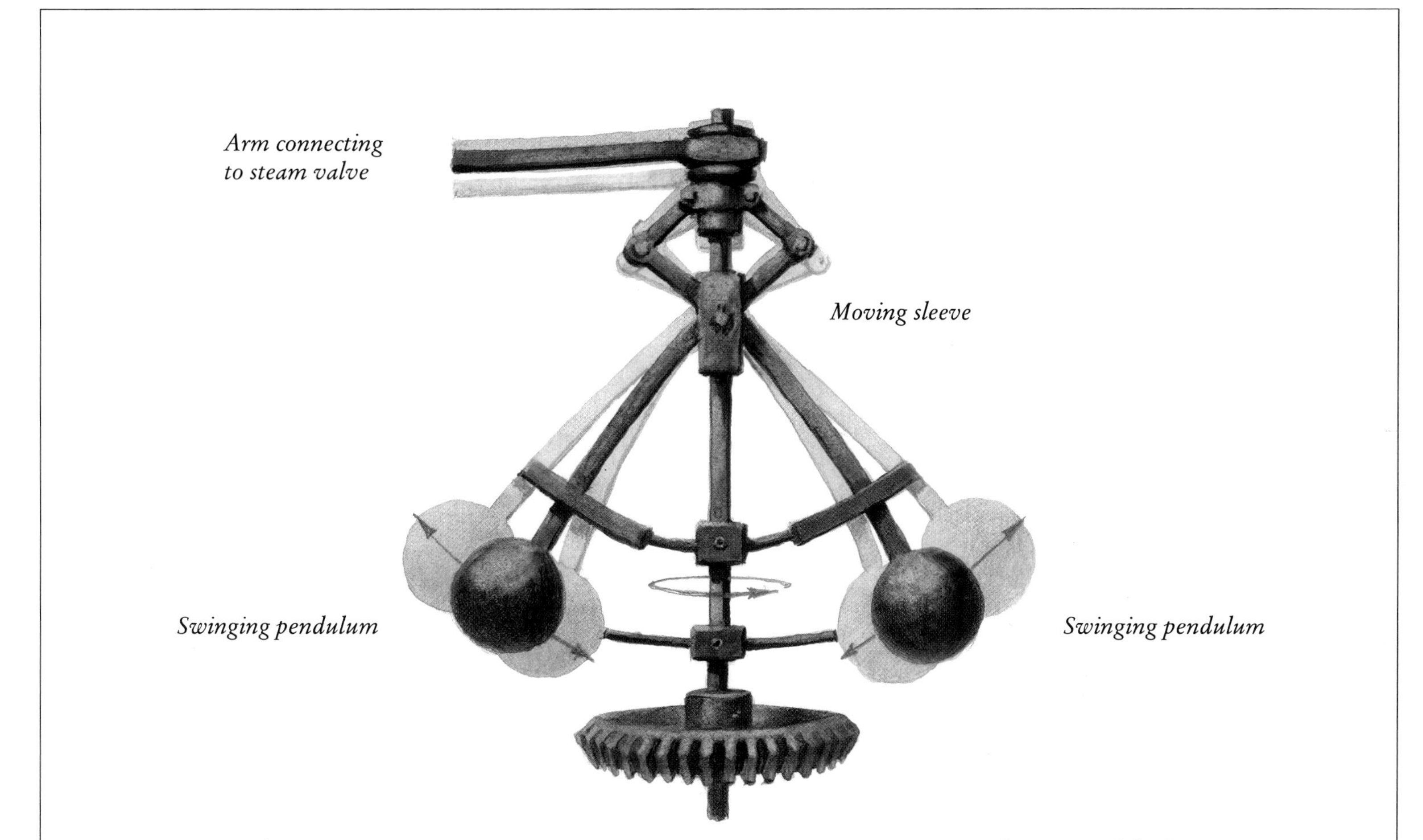

Automatic control

One of Watt's most important improvements to the steam engine was the introduction of the governor, a feedback device that allowed the engine to regulate its speed automatically. The governor was designed to cope with the fact that the speed of the engine tended to increase when the workload decreased, and go down when the workload went up. It consisted of two weighted pendulums, attached to a sleeve connected to the valve that admitted steam into the engine's cylinder. Because of the action of centrifugal force, the weights swung further and further outwards, the faster the engine ran. The outward motion of the weights lowered the sleeve, which rose as the weights moved inwards. The steam valve thus closed to let in less steam when the sleeve was lowered, conversely opening wider when the sleeve was pushed up, so the speed of the engine was kept as constant as possible.

The governor was not Watt's own invention. He probably got the idea from a device used to regulate the pressure of millstones, introduced in 1788. Watt applied it to the steam engine in the following year. But Watt was well aware of the importance of the governor. Although he could not patent someone else's idea, he tried to keep its use in his steam engines a secret for as long as possible, to steal a march on his competitors until the general patent on his steam engine ran out in 1800.

Other control devices

The governor was a mechanical control system. With the harnessing of electrical power, electrical control systems also became possible. One of the most useful was the electric relay, an electromagnet that opens or closes another circuit when its own circuit is modified in some way.

The relay has been used in many areas, but most notably in telecommunications. The complex telephone switchboards that started to appear at the end of the nineteenth century would not have been possible without such electrical control systems, and their development for this purpose led to many other uses for controlling machinery in industry. Today, however, the microcircuitry of the computer has taken over many of the roles of these mechanical and electrical control systems.

UTILIZING THE EARTH'S RESOURCES

To begin with, there was the power of human muscles. If you wanted to move a heavy load like a massive stone block for a building, the only way was to drag it along. The heavier the load, the more people you needed to move it. It was the same when it came to driving machinery. There was little point in setting up a large pair of millstones to grind your corn, if you could not summon up the strength to turn them.

Animals helped. But they had to be fed, cared for and harnessed. And only someone with the power or money to keep a large number of mules or oxen could shift the heaviest loads. So animal power, though useful when combined with devices like treadmills, and invaluable in the fields, had its limitations.

It is not surprising, then, that people began to try other power sources. Like other inventors before and after, they looked at their environment. The Europeans, for example, saw fast-flowing streams; people in western Asia felt the power of prevailing winds.

So water wheels and windmills established themselves in different forms, in different parts of the world. They were the key power plants for hundreds of years, driving machinery and making easier the production of everything, from cloth to paper, flour to pins. They are still doing sterling service in both developed and developing countries today. Indeed, wind power shows no sign of losing its importance.

But the modern age has seen even more far-reaching discoveries and innovations in the use of resources. The world of manufacturing was transformed in the eighteenth and nineteenth centuries by the steam engine. Meanwhile, scientists like Volta and Faraday were working on electricity, the resource that would make possible still greater transformations in the twentieth century. Such developments have continued, with revolutions in power for transport ranging from internal combustion engines to rockets and jets, and with the coming of nuclear power generation.

From the outside, these innovations seem remarkably diverse. What does a light bulb have in common with a steam ship or a rocket? And yet many of them are closer than they seem, and each relies on many technologies that appeared before it. For example, power is still generated by water, even though a modern hydro-electric station looks very different from a medieval water wheel. Knowledge of the way steam can be harnessed in the steam engine helped the early designers of turbines, still used in power generation. And the principle of the piston is used in both steam and internal combustion engines.

It is common today to hear wind-based power stations praised for their ecological soundness – a fair comment – and to hear people talking about a reversion to former technologies. Such developments are more typical than they might at first seem. The innovator is always looking both back and forward, and bringing together technologies to make a new whole.

NATURAL POWER

Harnessing the forces of nature helped to increase productivity, liberating people and animals in the process

To our earliest ancestors, power to do physical work meant the power of human muscles. If you wanted to move a heavy stone to make a house or a temple, you dragged it yourself, perhaps with the help of neighbours and a sled. If you wanted to grind corn to make bread, you ground it yourself, using a simple stone quern and the power of your own arm muscles.

The situation was similar in early civilizations, such as those of Egypt, Greece and Rome. Here, larger building projects and public works were carried out with the increased power of a larger workforce – usually a work-force of slaves. Sometimes it was simply a matter of multiplying the number of workers to increase the power. Sometimes the mechanical ingenuity of the ancient world was combined with the muscles of the slaves to produce a powerful machine. A case in point is the treadmill; this is the open wheel, which rotated as slaves trod continuously inside it. This wheel could be made to power machinery, such as millstones, used to grind corn, or simple cranes to haul heavy stones up the side of a new building.

Most of the early civilizations had no need to supplement this human muscle power with any other source. They had an ample supply of labour – often slave labour – from conquered or dependent territories, as well as from local areas.

If there was a temporary labour shortage, animals could be used instead. Animals, however, were not always as successful as humans. People had learnt how to harness and use oxen from early on in the history of farming. But, although they were useful for pumping water for irrigation, these slow, heavy beasts were no good in treadmills or workshops. The alternatives were the horse and donkey. But the ox-harness did not adapt well to smaller animals – it could choke a horse if the beast pulled too hard. So, when horses were used they could pull at only a fraction of their capacity, and it was just as effective to use humans.

The result was that human muscle power was the principle prime mover in the ancient world. Only much later were effective harnesses, stirrups, and horseshoes introduced to make the horse the accepted beast of burden that it became in the Middle Ages, and remains in some parts of the world to this day.

The coming of the water-mill

In the first century BC a development took place that would eventually change this situation. Someone thought of harnessing water power. The first hints and descriptions come from the Roman Empire. Water wheels are mentioned in a poem by the writer Antipater of Thessalonica (of the first century BC) and the Greek geographer Strabo (*c.*64 BC to *c.* AD 21).

The earliest form seems to have been a simple, wooden, vertical shaft with wooden scoops, or vanes, sticking out of its lower end. At the top the shaft passed through the middle of the lower millstone and was attached to the upper millstone. The lower end would be placed in a river, so that the flow of the water turned the vanes and the shaft, making the upper millstone turn at the same rate. This simple form of water-mill is often called the Greek mill.

The Greek mill worked, but it was not all that efficient. It needed a river, preferably a fast-flowing one, since the shaft and millstone turned at the same rate. It also needed a regular flow of water – there was no way of making the stone turn if the flow subsided for some reason. The device was taken up in some areas, but there was little need for it in the early, prosperous years of the Roman Empire, when the ruling classes had plenty of slaves at their disposal.

The development of the water mill

However, this fact did not stop Roman engineers improving upon this very basic design. The Roman architect and military engineer Vitruvius, also of the first century BC, was the inventor who developed water-mill design. He realized that the device could be made more efficient if some of the Classical world's knowledge of mechanics was applied to it. So he turned the wheel through ninety degrees and connected it to the millstone using gears. This straightaway gave it more potential for efficiency (albeit at the expense of turning force) because the gear ratios could be arranged to make the stones turn as many as five times for one turn of the wheel.

The first vertical wheels were under-

The noria

One of the earliest types of water wheel is the noria. This is a device with jars as well as paddles, which was used for raising water from one level to another. It was probably used in the Mediterranean area before the expansion of the Roman Empire.

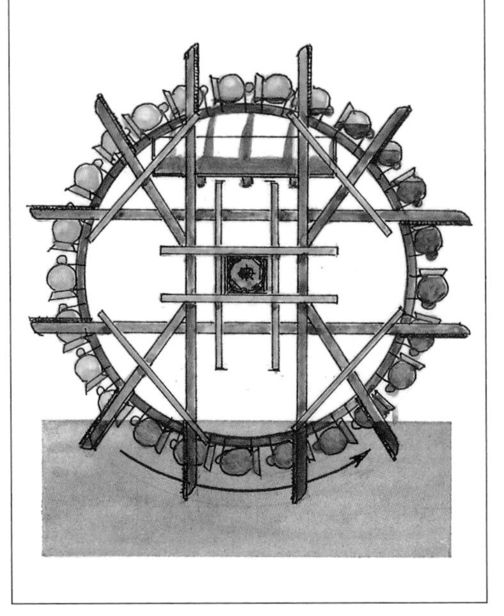

shot wheels, that is to say the water flowed straight under the vanes. Later it was realized that the vertical wheel could be improved by making the water flow over it in different ways. Thus the breast wheel and, most efficient of all, the overshot wheel were evolved.

Even the undershot wheel provided a far more efficient source of power than human muscles. It has been calculated that an undershot wheel excavated at Venafro, near Cassino, Italy, could grind 330lb (150kg) of corn per hour, more than twenty times that ground in the same time by two slaves using hand querns. It was an enormous advance.

An invention waiting for its time

Water-mills were mentioned as far afield as Denmark and China at around this time, yet the Romans did not take up the water-mill in large numbers until some 300 years after its invention. This seems strange: today, its advantages are obvious. The Romans had the technology to produce such mills, and the communications system to convey information about them around their empire. They needed flour to feed a large population and could have found many uses for water power.

There were problems of variable water flow in rivers, which could make water wheels unpredictable, but the advantages still look great. But, for most of their history, the Romans had enough labour not to need mechanization. Only when the expansion of the Empire had slowed down, and the effects of the reduced supply of labour began to be felt did the Romans seize on water power. Ironically, the decline of the Roman Empire led to the rise of this powerful new technology.

The Middle Ages saw more and more water-mills. For one thing, there was less of a concentration of slave labour. For another, individuals realized that they could make mills into profitable monopolies. Thus water-mills became part of the feudal system. A lord would have a mill on

Three types of water wheel

After the vertical water wheel had been established, three basic designs became common. The simple, undershot wheel was described by Vitruvius. The later breast wheel and overshot wheels were attempts to make the water wheel more efficient.

Undershot wheel

In this type, the water flows straight under the wheel, catching on the paddles as it passes. This is the least efficient type, but it requires a minimum of work to build and was therefore found useful in places where there was a fast, steady flow of water.

Overshot wheel

By diverting the flow so that it hits the wheel at the top and turns it by its descending weight, a more efficient harnessing of power was obtained. The paddles were redesigned as buckets to hold the water as it descended.

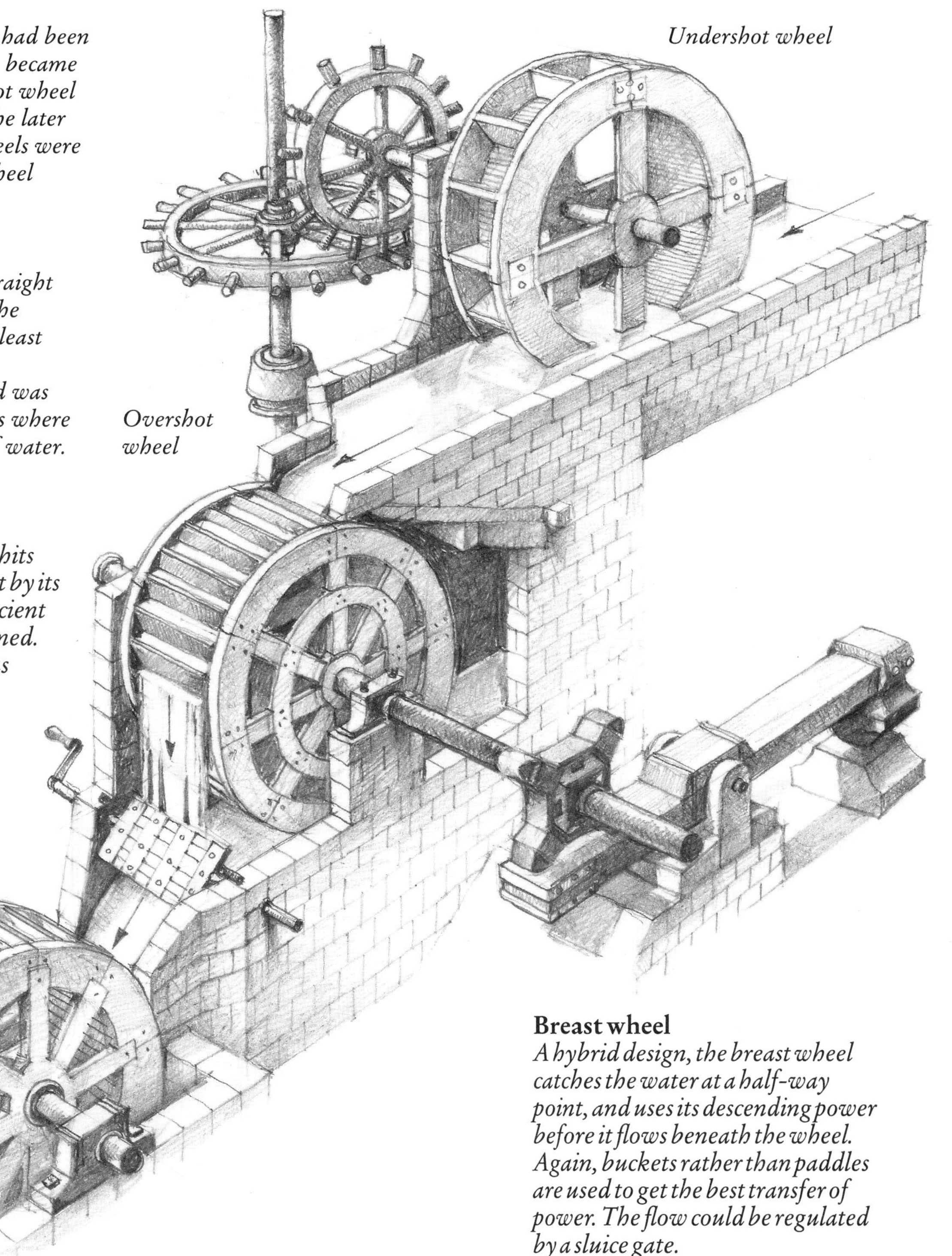

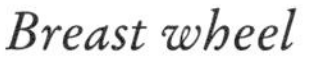

Breast wheel

A hybrid design, the breast wheel catches the water at a half-way point, and uses its descending power before it flows beneath the wheel. Again, buckets rather than paddles are used to get the best transfer of power. The flow could be regulated by a sluice gate.

his estate that he could hire out to the peasants in return for payment. Since no peasant could afford his own mill, and was in any case under the subjugation of his lord, he was obliged to grind his corn in this way. Monasteries also had mills, which the monks used to grind grain, process cloth and tan leather.

This fact is reflected in the figures. In the Aube district of France, for example, there were only fourteen mills in the eleventh century; by the early thirteenth century there were 200. The technology also began to spread around Europe. By the beginning of the thirteenth century water-mills had been set up in Iceland and Poland. We know less about their use in the Far East. They were certainly known in China long before their first recorded use in the West, and water power was being used for blast-furnace bellows in China in AD 31, an application that did not appear in the West until much later.

Uses and adaptations

Most of these mills were small-scale affairs. However, the Romans had already shown what could be achieved when the water wheel was applied to industry in a systematic way. For example, at Barbegal (near Arles, France) in AD 308, they created a stepladder arrangement of water wheels. Altogether, there were sixteen wheels, in two pairs of eight, along a river. These wheels used the same downward flow of water, and each drove two pairs of millstones. Twenty-eight tons of flour could be produced in a day using this early 'factory'. This capacity could have fed a population of around 80,000, although there were only 10,000 people in the Arles region at this time: the Romans probably used the extra capacity of the mill to feed a large section of their army.

It was not only in the production of flour that the water-mill became important. At the end of the Roman period and in the Middle Ages a host of other industries exploited this source of power. Sawmills for cutting wood and stone, paper mills, power-driven hammers for metalworking, mining hoists, grinding stones, tanning mills for the leather industry, and fulling mills (to make the cloth heavier and more compact) for the wool

From animal muscle power, to the forces of water and wind – natural power sources dominated the rural landscape for hundreds of years. The inset shows the gearing needed to turn the motion of a water wheel through ninety degrees.

trade were all being driven by water during the Middle Ages. There were mills for preparing malt and grinding pigments, for lifting water and for turning wood. Such a multiplicity of uses was sometimes found in a single town or city. In Bologna, for example, by the sixteenth century there were mills for grinding corn, for making copper pots and weapons, for spinning, for sawing, for sharpening and for polishing. Directly or indirectly, the impact of the water-mill affected nearly everyone in the Middle Ages.

Wind power

Where there were rivers, especially fast-flowing ones, the water-mill was useful. But there are large areas of the world, particularly the plains of Europe and the Middle East, where rivers are few and far between. These became the territory of the windmill.

The origins of the windmill are obscure. Furthermore, it seems to have appeared long after the period of the Roman Empire, and so there are no convenient Classical writers to describe it for us. It seems to have begun in the East. Some writers look to Tibet for the origins

Traditional middle-eastern windmill

of the windmill. There wind-driven prayer cylinders have been suggested as the inspiration for the new source of power. A more likely place of origin is in the Islamic world, where it is reported in the seventh century AD that some ingenious Persians had made a mill that was driven by the wind.

Other Arab writers mentioned windmills in the dry, windy plains of the Middle East in the succeeding centuries, although we have to wait until the thirteenth century to get a detailed description of how they were built. This description is by the Syrian writer Al-Dimashqi, who described a windmill designed rather like a Greek water-mill, with a vertical shaft and horizontal sails driving millstones above. This sort of mill is still seen in some parts of the Middle East today, but the stones are likely now to be below the sails, and there will be stout walls directing the prevailing wind to the sails.

The spread of wind power

From Persia, this sort of mill spread through the Islamic world and to the Far East. And it is from here, rather than from Tibet, that China probably got her first windmills. But how did the windmill spread to the West? It used to be thought that the Crusaders brought it home with them; others have suggested that trade routes through Russia and Scandinavia provided the link.

The truth may be that the technology of wind power did not travel in this way at all. Western mills from the start had horizontal shafts and vertical sails. They were thus very different from those in the East. It is possible that they were invented quite independently of their Eastern counterparts, perhaps by analogy with the vertical water wheel, perhaps picking up a hint in the writings of technological pioneer Hero of Alexandria, perhaps as a result of a process of trial and error that has left no trace other than the finished product.

Whichever was the case, windmills were becoming common in northern Europe before Al-Dimashqi's description of them appeared in the Middle East. In the late twelfth century they began to appear in English documents. The earliest is probably one mentioned at Weedley, in Yorkshire, in 1185, which was being rented for eight shillings a year. By the end of the century, soldiers of the Third Crusade were building a Western-style windmill in Syria, which suggests that – if the crusades were responsible for diffusing the windmill – the movement was actually from West to East, rather than the other way round.

Wind power spread quickly in Europe. During the thirteenth century about 120 windmills were built in the area around Ypres in Belgium. This area, together with northern France and The Netherlands, was ideal territory for the windmill. And such was the increase in the new technology, that Pope Celestine III decreed in the 1190s that owners of windmills should pay a special tithe.

Post mill and tower mill

The earliest surviving picture of a Western windmill is in the illuminated manuscript of an English psalter in 1270. It shows a structure that would be widely familiar for centuries afterwards. The sails were mounted on one side of a shed-like timber structure that contained the millstones. This in turn was set on an upright post supported by diagonal braces. The whole mill could be turned on the post to face the prevailing wind. This design was known, for obvious reasons, as the post mill.

The post mill was a simple and effective solution to the problem of positioning the sails correctly when the wind changed. But it took a lot of strength to turn the whole mill with its heavy stones: often a long handle was provided to give extra leverage when turning the post mill, and the weight of this handle helped to counterbalance the weight of the sails on the other side of the mill. But it was still a cumbersome device to use, and soon another design appeared. This consisted of a brick or stone tower with a revolving cap on top. The sails were attached to the cap, which could be moved into the wind, while the heavy machinery and millstones were housed below in the tower and did not have to be moved. Later still the cap was fitted with a smaller sail or fantail at the opposite side from the main sails. This would catch the wind when it blew, automatically turning the cap into the prevailing wind. This device has been considered the first example of automatic control.

The tower mill was a more solid and permanent structure and was easier to use. In addition, its construction did not depend on a supply of long, heavy load-bearing timbers. Another advantage was that the design could be adapted easily for use on one of the towers of a castle. It was very useful for castles to have such a source of power on site, which could make them still more independent than usual in times of siege.

Like the water-mill, the windmill's first use was for grinding corn. But the Dutch, with their pressing need for drainage, adapted it in the fifteenth century for pumping water by setting up series of post mills to pump the deep water away from reclaimed land. This adaptation heralded other uses such as wind sawmills, though the windmill never proved quite as adaptable as its water-driven counterpart. Nevertheless, it provided as enduring a power source as the water-mill, in the flat regions of northern Europe and the arid plains of the Mediterranean in particular, where in many regions, it has been the main source of power for centuries.

The tower mill and post mill, with their different turning mechanisms, are shown in the line drawings. In the foreground is an Australian mill made by pioneers – an efficient use of available resources.

THE STEAM ENGINE

Steam had fascinated inventors since ancient times, but its successful harnessing involved many minds

There is a famous story that the young James Watt sat in his parents' Scottish kitchen, transfixed at the sight of steam coming from a boiling kettle making the lid jiggle, and in a flash of inspiration conceived the steam engine and began the Industrial Revolution. Of course, it is untrue. Watt did not invent the steam engine, but improved an invention that already existed. What is surprising is how, long before Watt, people had contemplated boiling vats and kettles and wondered about the power of steam.

The first to record his thoughts was the engineer Hero, who lived in Alexandria some time between the second century BC and the third century AD. Hero invented a device called the aeolipile, a spinning metal sphere powered by steam. It did not have any practical application, but it showed just how far back the interest in steam power went. Other theorists followed in Hero's footsteps. They included men of the Renaissance, such as Giovanni Branca (1571–1640), who designed an unsuccessful steam-powered device for crushing metal ore before it was smelted.

The advance of theory

People like Hero of Alexandria and Giovanni Branca were working in the dark, without really knowing what steam was – or even understanding the essential properties of a gas. Such knowledge was essential before a working machine could be built to harness the power of steam.

One of the first people to grasp the properties of a gas was Frenchman Denis Papin, who did important work on gases and vacuums in the late seventeenth century, and is famous for inventing the pressure cooker. Papin had ideas for steam engines that would drive factory machinery, but his most famous design was for a steam engine to power the fountains on the estate of his patron, the Duke of Hesse. It worked, but pipes kept bursting because of the high pressure.

The pumping engineers

At about the same time an Englishman, Thomas Savery, was also trying out designs for a water pump powered by steam. Savery's work had an immediate practical application. Pit owners wanted to mine deeper seams as coal reserves near the surface ran out. But the problem with going deeper was that the shafts tended to fill up with water. So Savery was looking for a way to pump water out of coal mines. The tin miners of Cornwall had a similar problem with water, and it is no accident that both Savery and his successor, Thomas Newcomen, came from the English West Country.

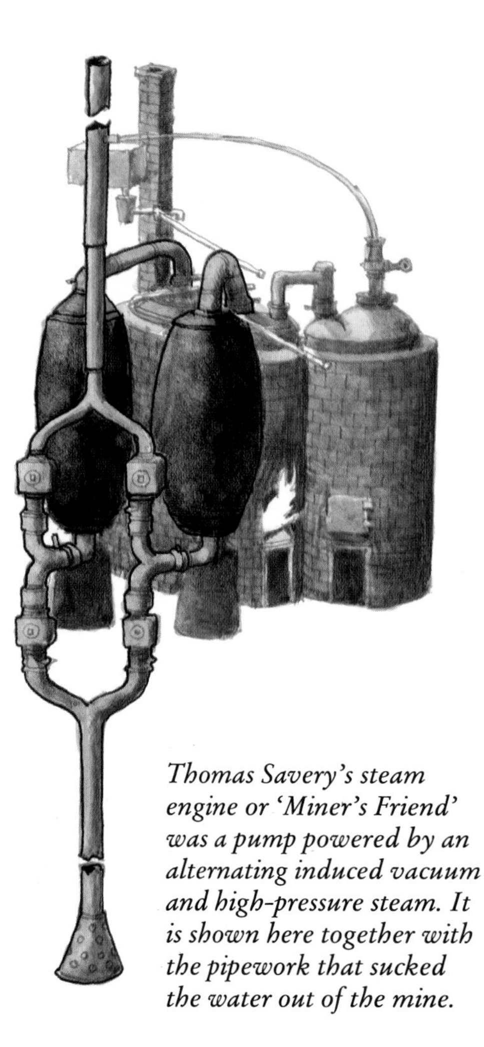

Thomas Savery's steam engine or 'Miner's Friend' was a pump powered by an alternating induced vacuum and high-pressure steam. It is shown here together with the pipework that sucked the water out of the mine.

In 1698 Savery patented his pump, the 'Miner's Friend'. It had a boiler that produced steam to fill a fifty-gallon (190-litre) container. When this was cooled (by pouring water on the outside), a partial vacuum was created inside, drawing water into the container from the mine shaft. Another injection of steam under pressure pushed the water further up the pipe. Savery's machine was limited: it would pump water up only about twenty feet (six metres). The fact that it worked at all was due as much to atmospheric pressure as to steam power. So it was used more successfully for pumping water in large houses than out of deep mines.

The engineer who took the steam engine the vital next step forward was Thomas Newcomen. In his engine the cylinder was cooled by a spray of water on the inside, and a piston was added to make it a true reciprocating engine. It was slow, heavy, and highly inefficient. But it was built solidly, and with as much precision as the metalworkers of the time could manage, and it was safer because it worked at a lower steam pressure. Its high coal consumption meant that the tin miners of Cornwall – for whom it was first intended – found it impractical: they would have had to bring in vast quantities of coal by sea to feed it. But coal miners took to the Newcomen engine and during the eighteenth century it became a common sight in northern England.

Watt's improvements

The work of Newcomen alone would have spurred on the Industrial Revolution. His engines allowed more coal to be mined from deeper seams, and this extra fuel was the key to extra industrial capacity. But the work of James Watt, a maker of mathematical instruments at the University of Glasgow, would make the engine still more influential.

Watt, influenced no doubt by the analytical approach of the Scottish Enlight-

Precursors of the steam engine

From the ancient world to the Renaissance, people were fascinated with the potential power of steam, but not sure how to harness it. For them it was a curiosity with apparently little practical use.

Hero of Alexandria's aeolipile consisted of a container that could be filled with water and heated. This was linked by pipes to an upper sphere with two protruding jets. Steam from the lower vessel rose up the pipes into the sphere and out of the jets, making the sphere rotate by recoil.

Giovanni Branca's ore-crusher had an obvious debt to the water wheel. Steam from a heated vessel was forced through a narrow pipe on to the vanes of a turbine that looked very much like a horizontal water wheel. This was connected by gears to a mechanism that was supposed to crush the ore, but the boiler would not have created enough energy to do the work required. Another Renaissance engineer, Salomon de Caus, drew on Hero's work to use steam to power the jets of fountains. His machines actually worked – but he did not exploit them in an industrial context.

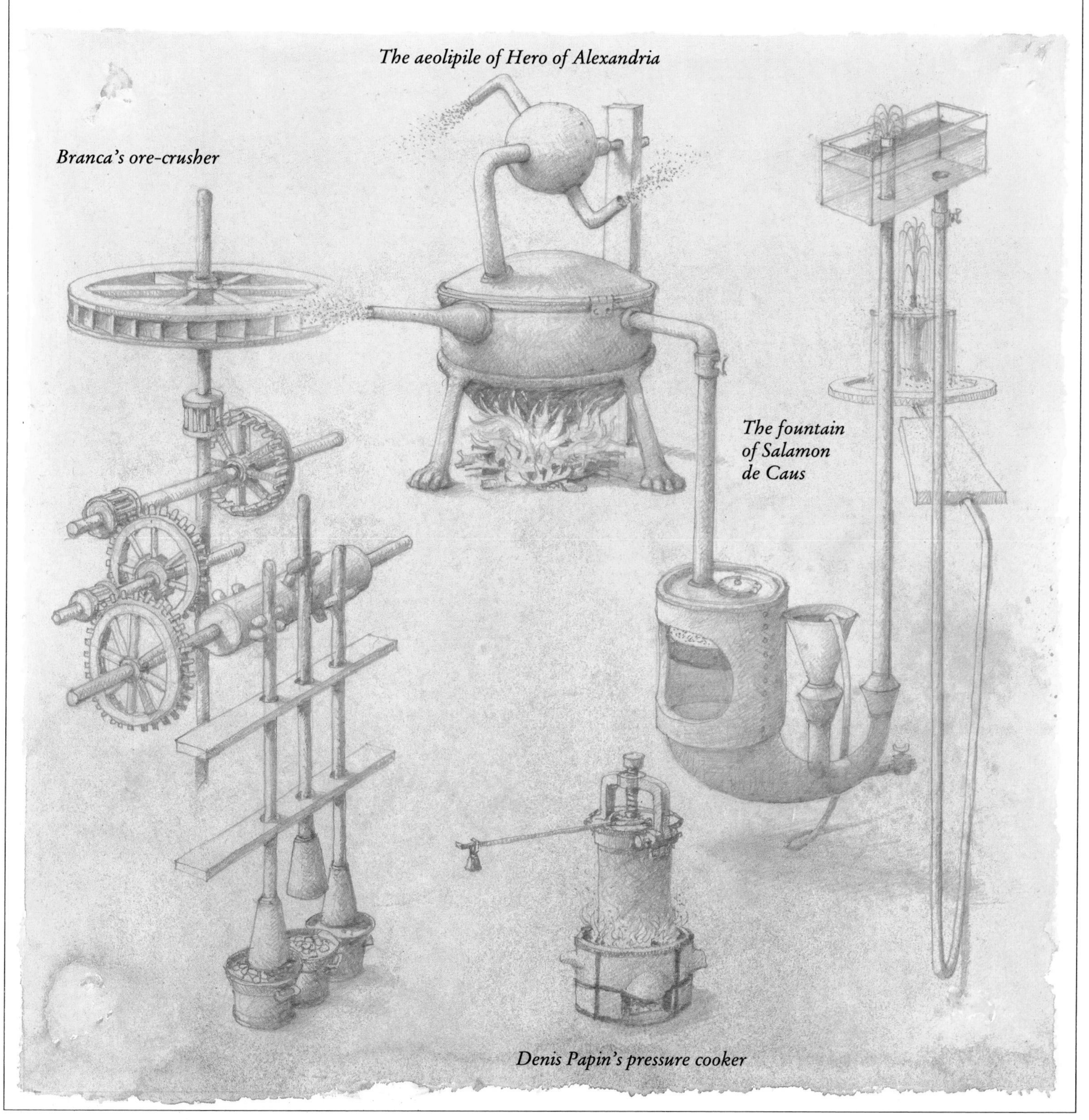

enment, saw clearly what was wrong with the steam engine as it stood. For a start, the cylinder had to be alternately heated and cooled with every stroke as the hot steam was admitted, and then condensed by cold water. Watt designed a separate condenser in which the latter process could take place. The condenser was kept continuously cool, the cylinder continuously hot. This modification alone saved much fuel. In addition, it made the engine faster because there was no pause while the cylinder reheated. Further efficiency was achieved by making the engine double-acting – that is to say, allowing steam to enter alternately, on both sides of the piston. Another refinement was the introduction of the governor, which controlled the speed of the engine by continuously regulating the throttle valve. The result of all this was an engine two to three times cheaper to run than Newcomen's.

In parallel to these improvements, Watt was working on ways to convert the up-and-down nodding motion of the cylinder into a rotary movement to turn a wheel, or drive machinery. He began with a mechanism based on a crank, but when one of his staff stole the idea, Watt came up with something more revolutionary – the sun-and-planet gear. In this arrangement, the engine's rod was connected to one gear, the 'planet', which 'orbited' a second gear, 'the sun', connected to the wheel shaft, making the latter turn.

A spreading influence

By 1787 all these improvements had been applied to the engine, and Watt was in partnership with manufacturer Matthew Boulton. The Boulton and Watt company could therefore offer mine-owners and industrialists an engine that was highly efficient and that could drive factory machinery. Suddenly, one did not need to be near a river with a water wheel to build a factory with power-driven machinery. The new prime mover could be built anywhere – and fuelled with the new supplies of cheap, deeply mined coal that the steam engine itself had made possible. The results, in terms of the spread of industrialization, the increase in production, and the social changes that went with it, were enormous.

Great Britain, the home of Boulton and Watt's factory, was the first to experience these changes – by 1800 Boulton and Watt had already supplied 500 steam engines to add to the many Newcomen engines already in service. This expansion of steam power coincided with the build-up of an empire that opened up new sources of raw materials and, to a certain extent, new markets for industrial goods. Soon, the power of steam was spreading elsewhere – back to the mainland of Europe where it had begun with the work of Papin, and to North America.

By the end of the eighteenth century another development was well underway. Pioneers like Watt, and Frenchman Nicolas Joseph Cugnot had started to experiment with wheeled vehicles driven by steam. Cugnot's steam carriage of 1770 was designed for hauling heavy artillery, and ran at a maximum of six miles (ten kilometres) per hour. In the early 1780s, Watt and his assistant William Murdock also developed a steam vehicle, which they patented in 1784. However, James Watt was not really interested in transport, and he prevented Murdock from developing the design further by giving him extra work on the stationary steam engine. It was left to different pioneers to adapt the steam engine for those other developments that were to fuel industrial and social change still more furiously – the railway and the steamboat.

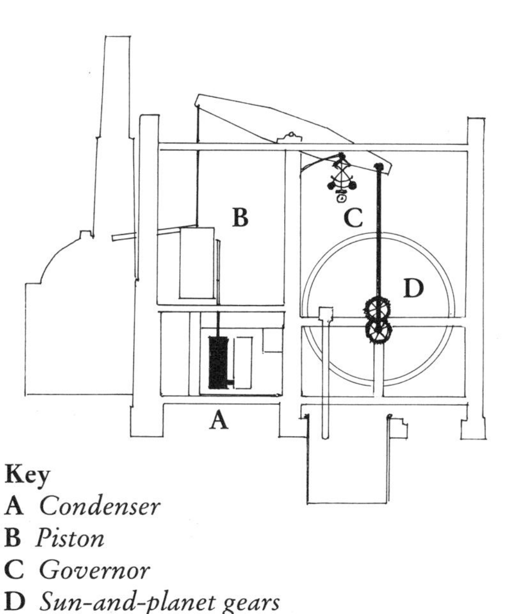

Key
A *Condenser*
B *Piston*
C *Governor*
D *Sun-and-planet gears*

Watt and the steam engine
James Watt made the following improvements to Newcomen's engine: the separate vessel to condense the steam; the double action, with steam entering the piston on either side; the governor to regulate the speed; and the sun-and-planet gear to convert up-and-down to rotary motion.

ELECTRICITY AND THE BATTERY

Convenience and safety in the home and workplace were much enhanced by the development of a clean and abundant energy source

Today most people take electricity's power for granted. Whether simply switching on the light or the radio, or taking advantage of hi-tech medical treatment, electricity is available, and we do not have to worry about it. In the developed world great networks of cables conduct electric current for miles, from power station to user; elsewhere, smaller generators (including domestic ones) are common and scarcely less reliable.

However, easily available electric power is a relatively recent development. As late as the early twentieth century, few people living outside the big cities had access to it. And even where there was domestic electricity, stories of power failures, and people using cushions to smother smouldering wires were all too widespread.

People knew about electricity long before this, though, because electrical charges exist as natural phenomena, a familiar example being lightning. This is a high-energy luminous electrical discharge that passes from a charged cloud to Earth. Anyone who has seen lightning damage can appreciate the tremendous power of electricity: the temperature it produces can be as high as 54,000°F (30,000°C). But for millennia no one could work out what caused it.

The same was true of static electricity. The Ancient Greeks noticed that when you rub a piece of amber, it attracts certain things for a while. But no one knew why this happened, or that there was any way of storing the power so produced.

As sometimes happens, the first steps were taken as much for entertainment as for practical reasons. Many an eighteenth-century amateur scientist experimented with a device called the Leyden jar, an early form of capacitor, which allows small charges to be stored.

Benjamin Franklin

As well as helping to draft the American Declaration of Independence and Constitution, Benjamin Franklin (1706–1790) carried out some of the most important early experiments with electricity. Fascinated by lightning, he attached a key to a kite by a thread of silk and flew the kite in a thunderstorm. Sparks leapt from the metal of the key, suggesting to Franklin that lightning was caused by discharges of static electricity. Connecting a kite to a Leyden jar and flying it in the same way produced a similar result – the jar was being electrically charged.

Benjamin Franklin

This led Franklin to realize that a pointed metal rod on top of a house, connected to the earth, could divert lightning's energy harmlessly to the ground. The lightning conductor was born. This beautifully simple idea has been saving lives and buildings ever since.

Franklin also came up with a cogent theory of electricity. He suggested that there were two types of electricity, one 'undercharged with electrical fire', the other 'overcharged'. Today, we still talk about negative and positive charges, although we know that it is electrons, not 'electrical fire' that are involved. Similarly, when Franklin proposed that when an 'overcharged' body approached an 'undercharged' one an electrical spark would leap between them, he was anticipating the modern concept of an electrical current of electrons passing between bodies of different potential.

Alessandro Volta

While enthusiastic amateurs were experimenting with Leyden jars, and Franklin was working out his theories, the Italian physicist Alessandro Volta (1745–1827) was taking some of the next important steps in the story of our relationship with electricity. Dissatisfied with the Leyden jar, he produced the electrophorus, another device for storing an electrical charge. This made use of a substance called ebonite, which was known to acquire a negative charge when rubbed with a dry cloth. In the electrophorous, a metal plate was held with an insulated handle above a charged piece of ebonite. The result was a positive charge in the lower surface of the plate, and a negative charge on the upper surface. The negative charge could be removed if a metal wire connected the upper surface to the earth, with the result that quite large quantities of positive charge could be stored.

The electrophorus was a success. It soon replaced the Leyden jar as a way of storing a charge. It also made Volta famous. But he was to go on to produce a

Italian physicist Alessandro Volta is shown against the background of some of his drawings of his experiments and inventions. The apparatus on the left is Volta's electrophorus, a device that could store a positive charge in large quantities. When the upper surface was placed on a negatively charged surface it acquired a negative charge, while the lower surface acquired a positive charge. Behind Volta's head is part of a letter he wrote announcing the invention of the battery. The columns of different metal discs separated by discs of cardboard can be seen clearly next to Volta's text.

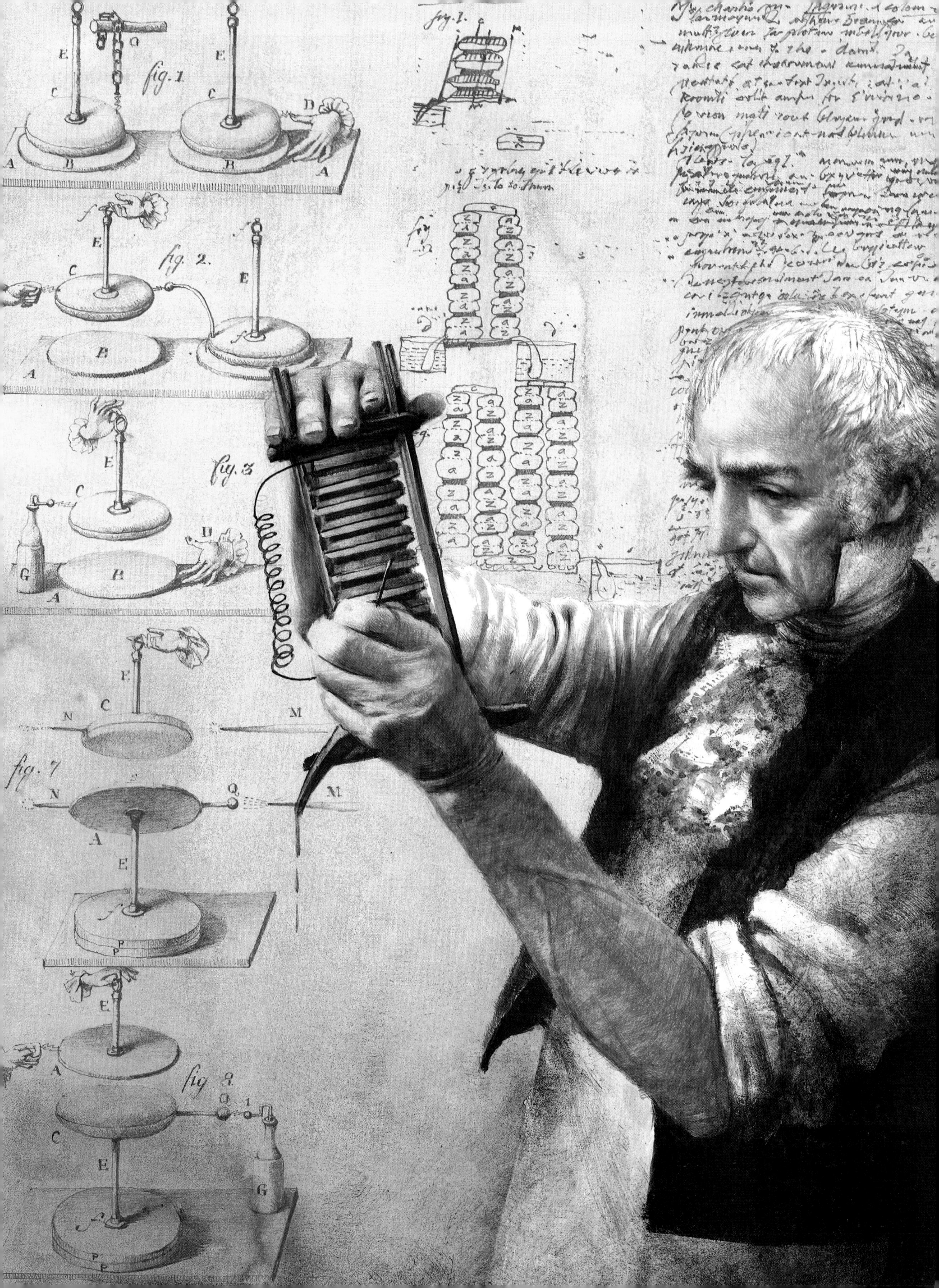

fig. 1
fig. 2
fig. 3
fig. 7
fig 8

yet more influential invention: a means of producing a continuous steady current – in other words, a battery. What he produced was a 'pile', consisting of a number of copper and zinc discs separated by pieces of paper soaked in acidulated water. There was a slight electric discharge at the ends of the pile. Another version of the experiment, made in 1800, was also successful. He part-immersed a sheet of copper and a sheet of zinc in a sulphuric acid solution. Then the ends of the sheets above the solution were connected with a conductor and a continuous electric current flowed between the sheets.

Volta's piles had to undergo decades of development before the emergence of the convenient and portable source of energy that is the battery we know today. But his work was important because it taught people some of the basic principles of electricity: how the copper and zinc plates act respectively as positive and negative electrodes, and how a difference of electric potential is set up between the two electrodes, which causes the charge to flow from one to the other.

The electric motor

Soon after Volta's experiments with batteries, other scientists began to explore different, but equally important aspects of electricity. In 1820 a Danish physicist called Hans Christian Ørsted discovered how a compass needle would turn when placed near a conductor carrying an electric current. This discovery was examined by the French physicist André-Marie Ampère who repeated Ørsted's experiment and formulated a rule to indicate in which direction the needle would turn. This was what he called the 'swimmer's rule' – if you imagined yourself swimming along the wire in the direction of the current and facing the needle, the north pole of the needle would point towards your left hand. Ampère's other rule says that if you grip the current-carrying wire with your right hand, and point your thumb in the direction of the current, your fingers will point in the direction of the needle's deflection.

The Danish physicist Hans Christian Ørsted was the first to demonstrate the relationship between electricity and magnetism and thus make possible the electric motor. Ørsted's experiment with a compass needle and a wire carrying an electric current set up vibrations through the whole world of physics. Scientists all over Europe repeated the experiment, while in the new world it inspired Joseph Henry to begin a series of experiments that would eventually lead to the invention of the electric telegraph. But it was Michael Faraday who did the most far-reaching work based on Ørsted's discovery, producing the first electric motor in 1821.

It was the British scientist Michael Faraday, however, who did the most far-reaching work based on Ørsted's discovery. Faraday realized that if he could isolate a magnetic pole (which cannot actually be done) it ought to move in a circle around a wire carrying a current. A device that could do this would be able to convert electrical energy into a potentially useful form of mechanical energy.

In 1821 Faraday came up with an experiment involving two beakers of mercury with an electric current passing through them. Both beakers contained a cylindrical bar magnet. When the current was flowing, one of the bar magnets, which was freely pivoted, rotated around the central current-carrying wire in its beaker. In the other beaker, where the magnet was fixed, the wire rotated around the magnet. In both cases, the rotation lasted as long as the current was flowing. In effect, Faraday had produced the first electric motor, an invention that would be as long-lasting as any from this age of the electrical pioneers.

Power generation

But Faraday's interest in electricity went further. In 1821 he used it to produce mechanical rotation. In 1822 he gave himself the opposite goal: to use magnetism to produce electricity. A key experiment involved winding five lengths of insulated copper wire around a ring of soft iron. Three were wound around one half of the ring, two around the other half, and one group of wires was not in contact with the other. The wires of the second group were connected to a galvanometer (an instrument for detecting small electric currents), and those of the first group were connected to a battery. When the connection was made to the battery, creating a magnetic force in the first group of coiled wires, the galvanometer's needle moved. Surprisingly, it also moved when the connection was broken.

Faraday next decided to introduce a permanent magnet into his experiments with coils. He discovered that introducing a permanent magnet into a coil induced a current in the coil. The current did not continue to be produced when the magnet was left stationary in the coil, but removing the magnet produced a current again – this time in the opposite direction. So if one continuously removed and inserted the bar magnet into the coil, an alternating current, running first one way

Michael Faraday, the English scientist who did some of the most important work on electricity and magnetism, is shown in front of sketches he made of some of his experiments. These include: 1) his repetition of Ørsted's experiment with the compass needle; 2) his demonstration of electromagnetic rotation; 3) a floating conductor revolves around a magnet, the principle of the electric motor; 4) the experiment with the coils of wire, his discovery of electrical induction; 5) his demonstration of the principle of the generator; 7) the principle of the dynamo.

P
N
1
2
N
3
C
5
4
B
6
N

then the other with each insertion and removal, would be produced. The phenomenon of electromagnetic induction was discovered independently by Joseph Henry, who was to become the first director of the Smithsonian Institution in Washington. Faraday was the first to publish his findings, however, and the law of electromagnetic induction is therefore called Faraday's Law.

Faraday's simple coil-and-bar magnet made up the first dynamo. It led Faraday to another important conclusion. If you could produce a current with a magnet moving in a magnetic field, you should also be able to produce a current by moving a conducting plate in a magnetic field. And so he arranged a disc of copper on a spindle and placed it between the poles of a powerful magnet. The spindle had a handle that allowed the operator to turn the disc. Faraday connected the spindle and the rim of the disc to a galvanometer, and while the apparatus was turning, a steady current was registered, this time a direct, not an alternating, current.

This research of Faraday's was of vital theoretical importance. But the British scientist did not develop the practical implications of the dynamo – for Faraday, it was interesting for what it told him about the properties of electricity, not for its practical applications. One man who did develop the dynamo further was French instrument-maker Hippolyte Pixii, who worked with André-Marie Ampère and took up Faraday's ideas. Pixii produced a generator that was much more efficient than Faraday's. It had two coils, beneath which a horseshoe magnet was turned by a hand crank. Pixii's generator also incorporated a commutator designed by Ampère. The latter made the electric current reverse itself, allowing a direct current to be taken from the generator.

Electric light

One of the things that stimulated Gramme and his contemporaries to develop the dynamo into an efficient, practical machine was the appearance of types of lighting that ran off electric power. The earliest of these was the arc lamp, which had first been seen when Humphry Davy, the scientist who took the young Faraday under his wing, was experimenting with Voltaic piles. Davy let two carbon electrodes touch and then drew them gradually apart. The electric current still flowed between them, in a brightly glowing arc. Arc lamps were being improved during the 1870s, and they were often used in factories and lighthouses, powered by Gramme dynamos.

But the great breakthrough in electric lighting came with the introduction of the carbon-filament lamp. Two men were developing the idea of a filament lamp at this time. The first, Joseph Wilson Swan, was working in England. The second, Thomas Alva Edison, was the doyen of American inventors. Both men knew that there were three key problems that had to be overcome: the right material for the glowing filament had to be found; a vacuum had to be created inside the bulb; and an appropriate source of electricity was required. Swan had experimented with various different materials for incandescent filaments since the late 1840s. Then, in 1875, William Crookes, working on his radiometer (a device for detecting and measuring radiant energy), announced a more successful way of removing the air from glass globes. By this time, Swan had hit on carbon as the best material for the filament, and he demonstrated a successful carbon filament lamp in 1878. Meanwhile, Edison came up with lamps using filaments of carbonized bamboo and cotton, which he announced the following year.

At first Swan and Edison worked in competition, filing patents and counter-patents in an attempt to protect their own ideas. Edison used his flair for publicity to great effect, lighting the streets of his home town of Menlo Park with the new bulbs. But in 1883 the two men combined their efforts to form the Edison and Swan United Electric Light Company. Since then electric light has spread rapidly across the world, closely following the availability of electric power.

Practical power generation

The nineteenth century saw many experiments with different types of generators, which, it was hoped, would generate electricity efficiently enough to do practical work. Generators had to develop in two main directions before they were used widely. First, instead of the permanent magnets in the early experimental generators, more powerful electromagnets were necessary. This allowed the production of a larger current from a smaller generator. Second, coil designs had to be modified, allowing the lines of force of the electromagnet to be cut more effectively.

These developments began to be realized in the 1860s, and in the following decade a truly practical dynamo for generating direct current was produced by Zénobe Théophile Gramme. Unlike many previous attempts, Gramme's dynamo did not overheat, although it required regular servicing. It entered service in factories and lighthouses, where the power it generated was usually used for lighting. Processes such as electroplating, which required a relatively low power, were also sometimes powered by Gramme dynamos, and the designer demonstrated his faith in them by equipping his own factory with them.

Better generators meant that a more constant electric current could be produced more cheaply than before. And so the developments in generator design were accompanied by developments in the design of those items that would use electricity. This meant that arc lamps were improved substantially, especially for street lighting. Where this type of street illumination could be provided more cheaply than gas lighting, it caught on quite quickly. In Paris, for example, arc lighting was being used widely by 1878. In New York, a successful demonstration of arc lighting on twenty city blocks of Broadway in 1881 also led to a rapid take-up of this type of illumination. Where gas was cheaper, the old technology proved more long-lasting. The same was true of incandescent lighting, also introduced in the early 1880s.

Expanding systems

Pioneering work with street lighting and in individual factories was impressive enough. But the effects of electricity could be still more far-reaching when power was supplied across a network to many individual subscribers. Thomas Alva Edison was one of the first to realize this, and set up his influential Pearl Street Station in the heart of New York City in 1882. In six years it was supplying some 65,000 lamps.

However, it was some years before the idea of a truly integrated electricity supply system took hold. One problem was the question of whether an AC or a DC system should be used. By the time of the 1893 Chicago World's Fair the two systems had been brought together by the American Westinghouse Corporation, who used the occasion to show their electricity generating system, which could supply both AC and DC at different voltages.

The Westinghouse system gained publicity when it was adopted for generating electricity at Niagara Falls. This electricity was required for a number of purposes, from supplying the industrial city of Buffalo, twenty miles (thirty-two kilometres) away from the Falls, to powering local railways, servicing factories, and fulfilling local lighting requirements – all at different voltages. It was an impressive project, and it gave the idea of an integrated electricity supply system publicity for the first time.

The Niagara project had one great advantage. It was a hydroelectric scheme: the force of the water from the falls was harnessed to drive turbines. But such potential for hydroelectricity existed in few places in the world. There were some similar schemes in the European Alps and in Scandinavia, but power had to be generated in other places by using large, heavy, vibrating reciprocating engines.

In London, in 1887, S. Z. de Ferranti began an even more ambitious project – a great power station at Deptford, south-east London, which would supply up to two million lamps using reciprocating engines to generate the power. It was not a financial success, but this was perhaps because the idea was ahead of its time: it set the trend for the structure of power generation in the twentieth century. Meanwhile, in Britain, small-scale generating stations continued to proliferate.

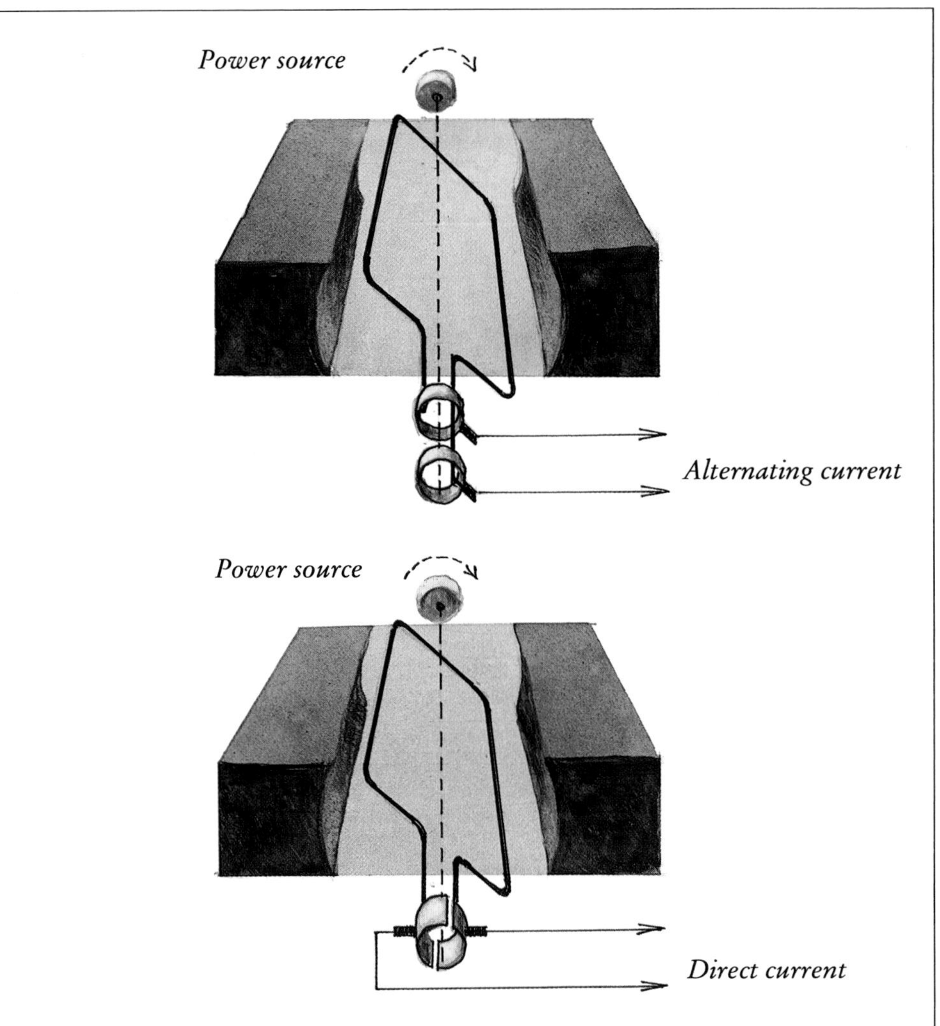

Two types of current

Whereas a battery produces direct current (DC) electricity, a dynamo, as we have seen, produces alternating current (AC) (although AC can be converted to DC with a commutator). The fact that there are two different types of current led to the multiplication of the number of generation and distribution systems that appeared at the end of the nineteenth century – and even to a battle between AC and DC. The problem with DC is that it is expensive to distribute over wide networks because a lot of energy is lost in transmissions. AC, on the other hand, tended to frighten people because of the high voltages involved.

Part of the problem was created by Edison, who had developed a DC distribution system in which he had a vested interest. Edison's company gave added publicity to the AC electrocution system developed by the rival Westinghouse company for the electric chair, only to sow the seeds of doubt in the minds of the public: could the supply used to execute criminals be safe for use by the rest of the population? It could, of course, and Edison was quietly ignoring the fact that many DC systems worked on voltages just as high as AC. But people felt safer, and DC was already established, so it was cheaper in the short run.

The change came when demand for electricity increased beyond that for lighting, particularly as industry realized the benefits that could be gained from electric motors. Croatian-born Nikola Tesla, who worked for Westinghouse, had developed the best-available electric motor, and it was an AC design. It ran at constant speed, unlike the DC motors that went before, and it was not prone to overheating and sparking, unlike its DC counterparts. In addition, American electrical engineer Charles Proteus Steinmetz played a key role in analysing how AC circuits behave, and in helping to bring about the dominance of power distribution by alternating current.

The work of Tesla and Steinmetz, combined with the power of Westinghouse, which developed an electricity distribution system capable of supplying both AC and DC at a range of different voltages, won the day.

INTERNAL COMBUSTION

Horseless carriages, originally playthings of the rich, created a transport revolution and had far-reaching effects on both trade and warfare

Today, we are all familiar with the internal combustion engine – the light, compact power source that drives motor vehicles all over the world. It is a unit in which the fuel is burned inside the engine, rather than in a separate furnace as in a steam engine. During the last hundred years it has brought relatively cheap, reliable transport to a large part of the globe – wherever there are roads of reasonable quality, and in many places where there are not. The revolution that replaced horse-drawn carriages with 'horseless' ones changed the world in many ways, making trade easier, transforming personal transport, bringing changes in every field from warfare to health care – and polluting the air.

All this has happened in less than a century. But the technology on which it depends has a longer history. Dutch inventor Christiaan Huygens experimented with an internal combustion engine in the seventeenth century. Huygens' engine used a small charge of gunpowder inside a cylinder to raise a piston. The piston moved down again under atmospheric pressure when the gases in the cylinder cooled.

The gas engine

After Huygens' engine, scientists concentrated on steam as the coming industrial power source, although the Dutchman's French follower, Denis Papin, performed further experiments with internal combustion. Internal combustion was thus largely ignored until it was taken up in the 1850s by French inventor Etienne Lenoir. He devised an engine in which the piston was moved by an exploding mixture of lighting gas (made from coal) and air, fired by an electric spark created inside the cylinder. The engine was double-acting. In other words, after the first explosion, as the piston started on its return stroke and the exhaust gases were expelled, a second explosion took place on the other side of the piston. The engine worked – but at a price: it consumed a large amount of gas in return for a small amount of work produced, and therefore did not supplant the steam engine.

France saw the next major development, which went a long way towards making this type of engine more efficient. Alphonse Beau de Rochas patented the four-stroke cycle in 1862. This cycle of operation was to be used in the majority of internal combustion engine designs that followed. It was the German engineer Nikolaus A. Otto who was to put this theory to work in a successful engine. His design of the late 1870s was highly effective, and many such engines were built for use in industry.

Methods of ignition

Igniting the fuel-and-air mixture inside the cylinder proved a major headache for early designers of internal combustion engines. Sometimes this was achieved with a flame, kept alight in a chamber next to the cylinder. At the moment of ignition a slide-valve opened a slot in the cylinder wall, drawing the flame into the cylinder and igniting the fuel. But the explosion blew the flame out each time, and a complex mechanism with a second flame had to be devised to re-light it before each explosion of the fuel.

A better method was hot-tube ignition. With this system, a porcelain or nickel tube was heated by a gas burner. At the moment of ignition a small section of the heated tube was let into the cylinder. A later variation was the open hot-tube ignition, in which the hot tube was actually connected to the compression chamber, and the exact time of ignition could be varied by moving the gas burner flame.

Today, petrol engines use an electrical spark, provided by a battery and a high-voltage induction coil, to create the explosion inside the cylinder. Diesel engines work by compression ignition: they require no spark plugs because the fuel ignites as a result of the high pressure inside the cylinder.

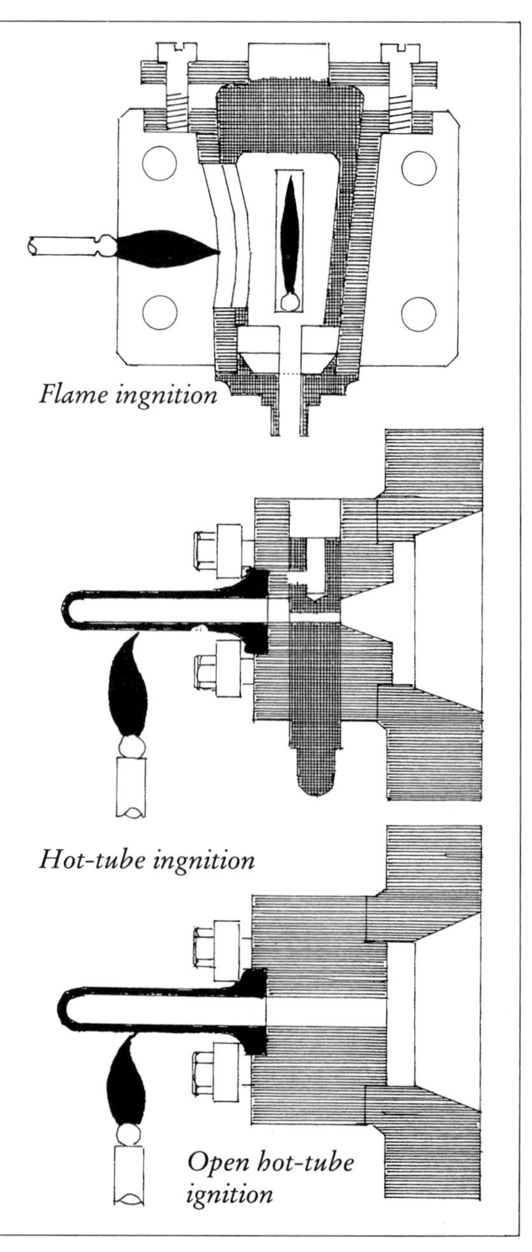

Flame ingnition

Hot-tube ingnition

Open hot-tube ignition

These early internal combustion engines were slow and heavy. They were seen as potential replacements for the steam engines that were used for pumping and driving machinery all over the industrialized world. There was little idea that such cumbersome devices might be used to power moving vehicles.

In any case, there were other problems in their design. Igniting the explosion inside the cylinder could be a problem. Some solved it with an electrical spark. Other engineers favoured introducing a flame into the cylinder at the correct point in the cycle. Even when the ignition difficulties had been overcome and, as in the later gas engines, the fuel consumption reduced, there was still the question of gas supply. For most industrialists, the steam engine continued to be the favoured source of power.

From oil to diesel

One solution was to use an oil such as paraffin as the fuel. In 1873, for example, both Brayton of Philadelphia and Hock of Vienna came up with such engines. These, together with other designs by the French engineer Capitaine, and by Dent and Priestman in England, continued to be developed in the later decades of the nineteenth century.

However, it was the engine that Rudolf Diesel patented in 1892 that was to have the more lasting influence. Diesel came up with the concept of ignition as a result of compression. Air was introduced into the cylinder and compressed, thereby increasing the temperature, so that when the fuel was forced into the cylinder it would ignite spontaneously. Diesel's engine worked on the same four-stroke cycle as Otto's gas engine. But pure air was first of all compressed by the moving piston, before an injector-pump forced the oil into the cylinder, where it ignited on contact with the compressed air.

The Diesel engine was a significant step forward – its efficiency was higher than any other engine in production at the time. The great bulk and weight of the early models did not seem to matter – after all, a stationary steam engine is also very large and heavy. So the Diesel engine caught on, and has remained an important prime mover to this day.

Petrol (gasoline) engines

In the meantime, researchers were developing other types of engine. The American George Brayton built a liquid-fuel engine in 1873. But perhaps the most influential developments were in a workshop in Württemberg, Germany, where an engineer was conceiving another engine. The engineer's name was Gottlieb Daimler, and his engine used petrol as a fuel. Daimler had much experience of engines using lighting gas as a fuel. He had been employed at the Deutz factory founded by gas-engine pioneers Otto and Lagen in the 1870s, when demand for gas engines started to increase rapidly. One of his colleagues there was Wilhelm Maybach, and the two men later set up together to develop the potential of internal combustion engines.

Daimler was interested in the idea of a small, high-speed engine that would be more movable and useful in a wider range of applications than were the gas and oil engines currently in existence. First of all, he and Maybach experimented with gas engines, but in the mid-1880s he started to work on an engine that would run on petroleum spirit. This vaporizes in the presence of air, so Daimler wanted to design a device that would bring the spirit and the air into contact with each other, thus making the spirit vaporize, and providing the volatile petrol-and-air mixture that would ignite in the cylinder.

In order to do this he created a surface carburettor, rather like the one used by Lenoir. Simply and briefly, this consisted of a metal container for the petroleum; air was drawn over the surface of the petroleum, picking up vapour on its way. There was a smaller container above the main vessel in which the carburetted air (the mixture of air and petroleum vapour) could be stored until it was needed in the cylinder. When the cylinder reached the point in its cycle when it needed the petrol-and-air mixture, this would be drawn in through a pipe.

Several experimental engines were produced by Daimler and it was the third of these that proved particularly successful. It could develop 0.5 horsepower and turned at 900 revolutions per minute (as opposed to the 200 rpm of the standard Otto engine of the day). What was more, it was small and light. The engineer connected it up to a bicycle in 1885 and tested the resulting vehicle triumphantly on the roads the following year.

It was a great success, but Gottlieb Daimler did not rest on his laurels. He continued to develop the internal combustion engine over the next decade or so. His 1889 engine had two cylinders in a V-formation with both cylinders connected to the same crank to turn their up-and-down movement into rotary motion. It was a simple and effective design, which would sell in large numbers, both for stationary and for mobile use; its light weight made it an obvious candidate for the latter. Indeed, Daimler had already harnessed

The four-stroke cycle

The four-stroke cycle, patented by Rochas and made practical by Otto, can be summarized as follows:

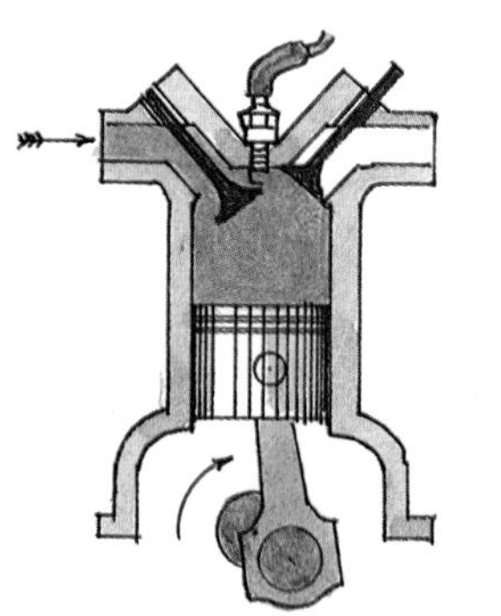

In the first stroke, called the induction or intake stroke, the piston moves down and the inlet valve is opened so that the explosive mixture of gas and air is drawn into the cylinder.

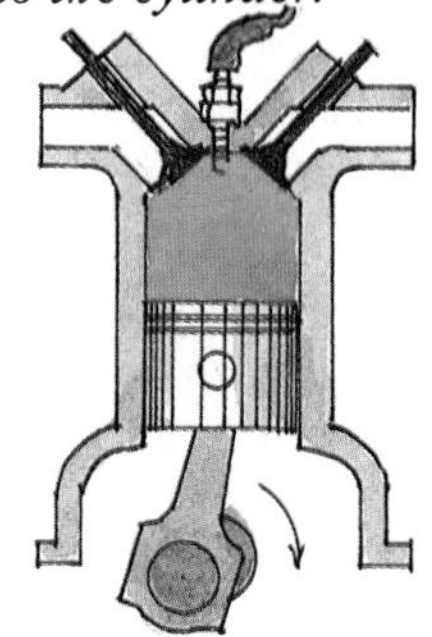

In the second stroke, the compression stroke, the valves are closed and the piston moves back; the explosive mixture is compressed. A spark ignites the fuel, and pressure increases sharply.

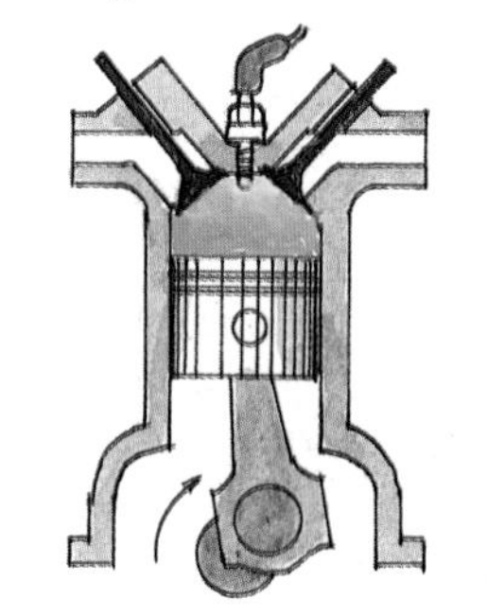

In the third stroke, the power stroke, the expanding gases in the cylinder force the piston back down.

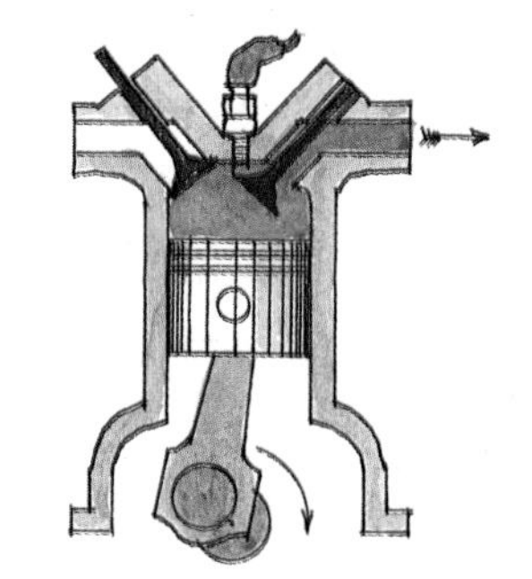

On the fourth stroke, the exhaust stroke, the burnt fuel is released from the engine through the exhaust valve.

one of his engines to a four-wheel vehicle in 1887 – but another pioneer had already created the first true motor car.

The coming of the motor car

Meanwhile, in Mannheim, Karl Benz was making his first experiments with powered transport. Benz was a manufacturer of gas engines. These were not remarkable in themselves. They used a two-stroke cycle (the four-stroke cycle of Otto had been patented) and were rather small. But their small size allowed their maker to adapt one for use on a two-seat tricycle. The single cylinder and flywheel fitted between the rear wheels. He fitted a carburettor to make it run on petrol, and by 1885 had the unlikely looking vehicle working. He patented his three-wheel car in 1886.

The car that shocked the residents of Mannheim in 1885 had an engine remarkably like one of Daimler's. But the two men had worked independently. Benz went on to develop further vehicles – this time with four wheels, and it was his four-wheeler design of 1893 that was to prove enduring. By this time Benz had actually sold his first car, to an enthusiast and businessman from Paris called Emile Roger. Benz granted Roger sole assembly and sales rights of his vehicles in France, and before long one could buy a Benz in Paris or Mannheim. And people did. Several hundred of the vehicles were built between 1893 and 1901 as the idea of the motor vehicle began to find favour amongst those who had enough spirit of adventure, and enough money, to try one for themselves.

To begin with, it was hardly a transport revolution. The cars were handmade and expensive, and so access to the new form of transport was restricted. Speed was limited too. A top speed of fourteen miles (about twenty-two kilometres) per hour was all that one could expect, and the early models often had difficulty getting up steep hills at all. But a radical change had occurred in the world of transport. By the time the effects of mass production were felt on the automobile industry, the revolution would be complete.

The coming of the motor car

Otto gas engine
In 1878 Nikolaus Otto produced an internal combustion engine that ran on lighting gas and air. Early internal combustion engines like this were too heavy to be used to power moving vehicles, but they were useful in industry.

Karl Benz

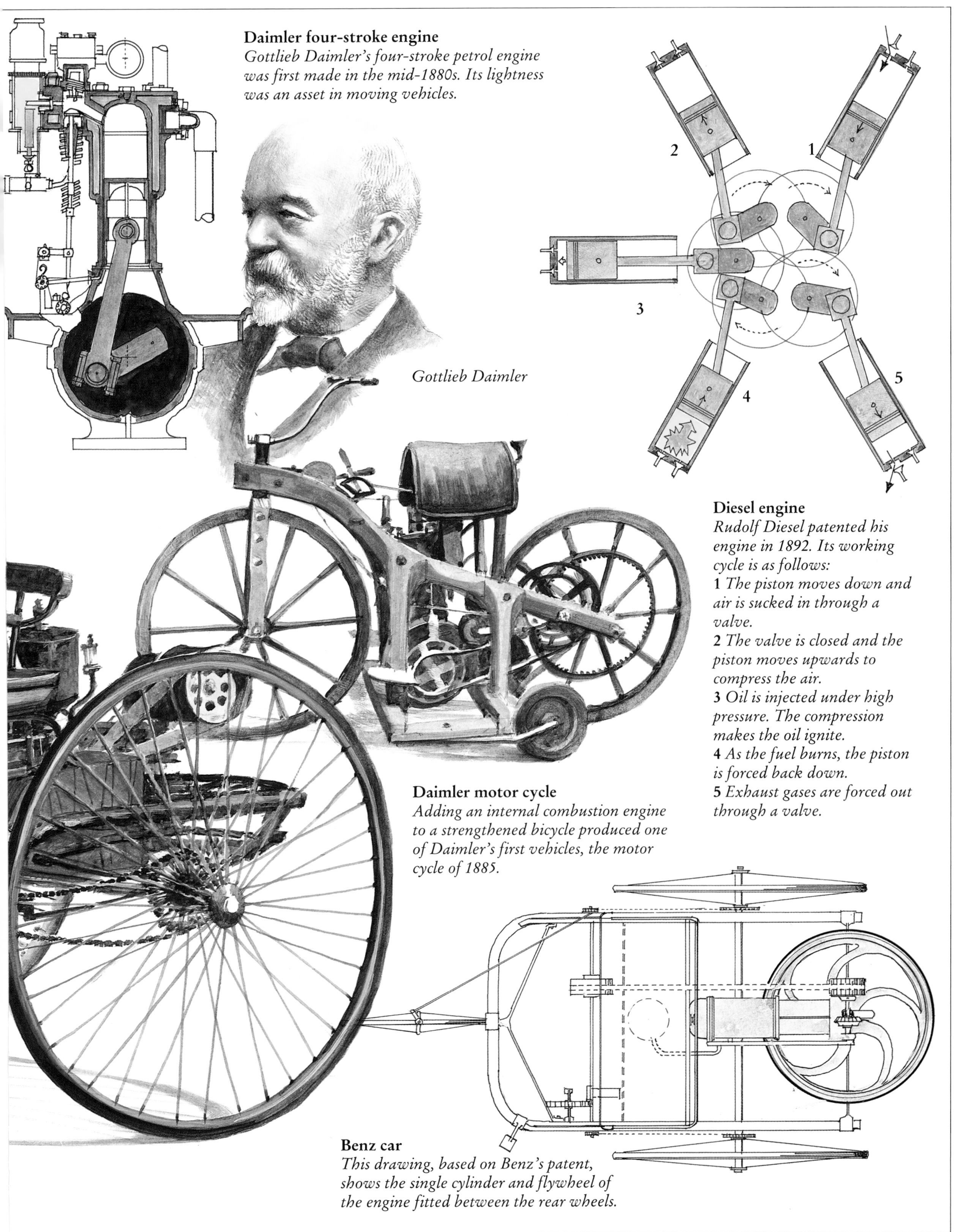

Daimler four-stroke engine
Gottlieb Daimler's four-stroke petrol engine was first made in the mid-1880s. Its lightness was an asset in moving vehicles.

Gottlieb Daimler

Diesel engine
Rudolf Diesel patented his engine in 1892. Its working cycle is as follows:
1 *The piston moves down and air is sucked in through a valve.*
2 *The valve is closed and the piston moves upwards to compress the air.*
3 *Oil is injected under high pressure. The compression makes the oil ignite.*
4 *As the fuel burns, the piston is forced back down.*
5 *Exhaust gases are forced out through a valve.*

Daimler motor cycle
Adding an internal combustion engine to a strengthened bicycle produced one of Daimler's first vehicles, the motor cycle of 1885.

Benz car
This drawing, based on Benz's patent, shows the single cylinder and flywheel of the engine fitted between the rear wheels.

Jet engines are used in most types of aircraft today, including modern jet fighters like this.

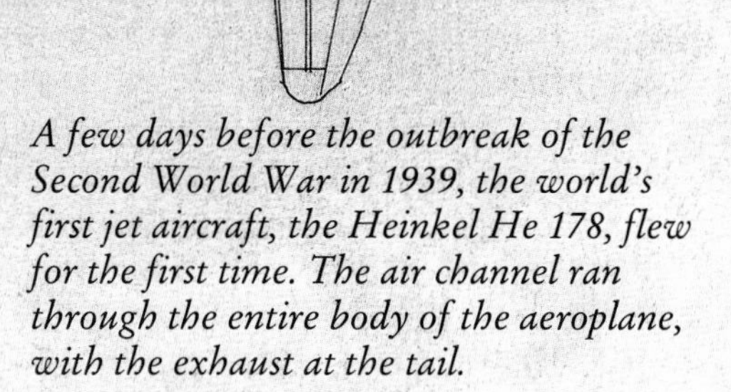

A few days before the outbreak of the Second World War in 1939, the world's first jet aircraft, the Heinkel He 178, flew for the first time. The air channel ran through the entire body of the aeroplane, with the exhaust at the tail.

Below: *Modern turbo-fan engine used mainly for powering civil aircraft.*
1 *Fans push air around the combustion chamber and into the tail pipe, providing extra thrust by air displacement.*
2 *Compressor forces air into the combustion chamber.*

Combustion chamber

Air intake

Rotary compressor

1

2

Air intake

indeed, the faster ones had to, to avoid the greater frictional drag of the atmosphere at lower altitudes. This meant that one could fly above the main weather systems, making flights smoother and pleasanter than before. Suddenly, inter-continental air travel was comfortable, quick and easy: it became normal to travel all around the world by air.

Breaking the sound barrier

Another area in which civil aviation eventually benefited from a military development was supersonic flight. The sound barrier had been broken by a Bell X-1 rocket plane in 1947. Supersonic bombers appeared in the 1950s. And at the end of the 1960s came the supersonic airliners, first the Russian Tupolev 14 and then Concorde, fruit of a collaboration between Great Britain and France.

But such freedom was not, and is not, cheap. Some of the early pioneers were put off jet engines and turbines because of the large amounts of fuel they consumed. They still do. A large-bodied aircraft needs about 42,000 gallons (190,000 litres) of fuel for a transatlantic crossing. This vast quantity of fuel has to be found and paid for. It also has to be housed safely during the flight. The jet engine could be said to be a typical symptom of the consumer society of the second half of the twentieth century, a blessing and a curse combined.

Frank Whittle's engine was used in the Gloster E 28/29 aircraft. The first flight with this engine was made in May 1941.

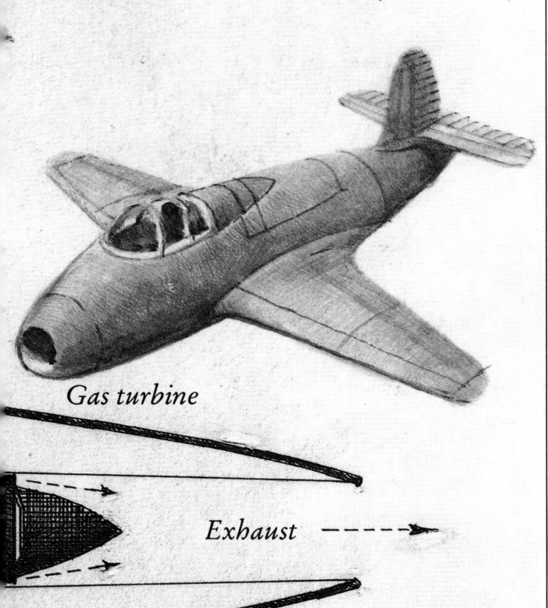

Gas turbine

Cross-section of Whittle's engine, as fitted to the Gloster E 28/29

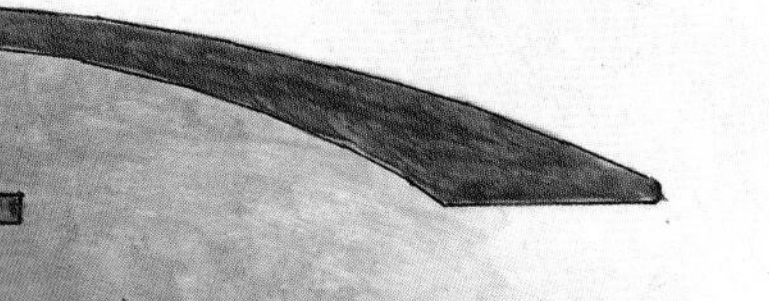

Fuel enters the engine, mixes with compressed air and burns in the combustion chamber.

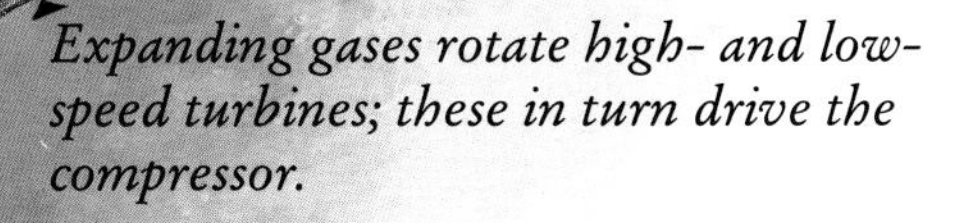

Expanding gases rotate high- and low-speed turbines; these in turn drive the compressor.

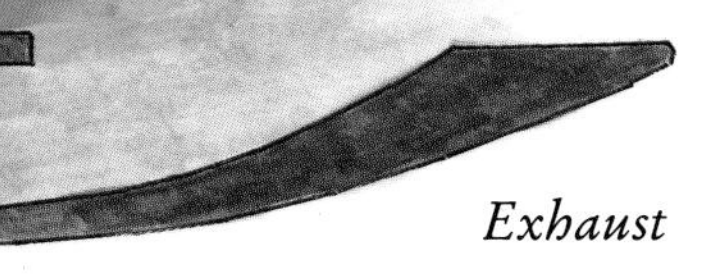

Types of jet engine

There are several types of jet engine. They all rely on the same basic principle, creating thrust from their exhaust gases, and burning a mixture of kerosene (paraffin), and air from the atmosphere. The turbo-jet is best suited to fast flight and is most commonly used in fighter aircraft. The turbo-prop is designed for lower-speed flight and is widely used in civil aircraft. So is the turbo-fan, which is valued because it is quieter and uses less fuel than the other types of engine.

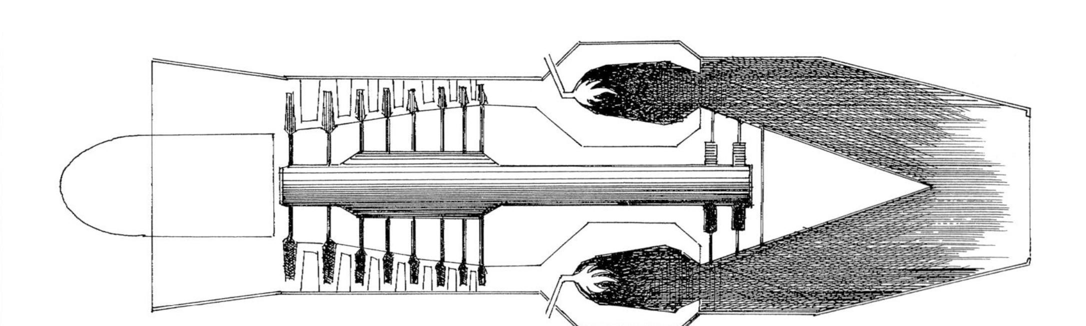

The turbo-jet (above) *is the simplest form of jet engine. In this type all the thrust comes from the exhaust gases as they leave the rear of the engine. There is a turbine, but this is used to drive the compressor only.*

In a turbo-prop engine (below) *there are two turbines. One (a low-pressure turbine) drives the compressor. The other (at higher pressure) drives a propeller at the front of the engine. Most of the thrust comes from the propeller.*

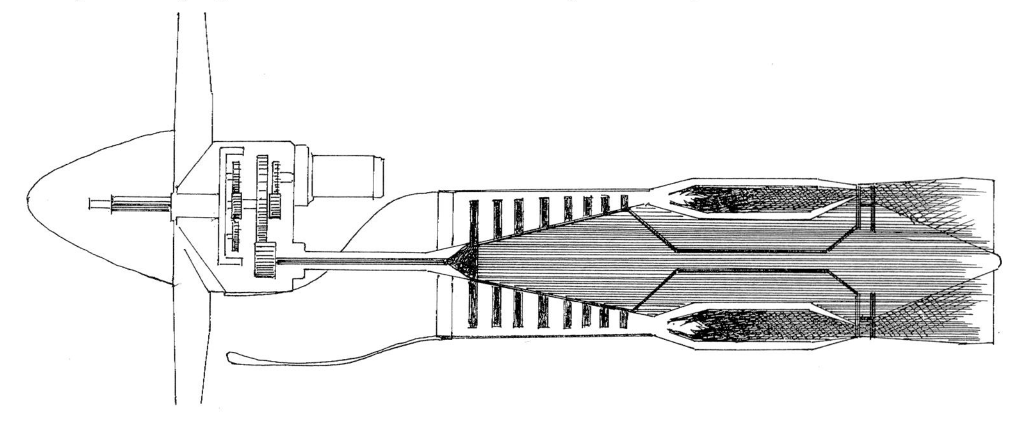

ROCKETS

Since the Chinese invented fireworks over 800 years ago, there has been a quest to invent devices that go higher, further and faster

A rocket works similarly to a jet-powered aircraft, moving forward as a reaction to the backward ejection of exhaust gases from the engine. But a jet engine relies on a supply of oxygen from the Earth's atmosphere to mix with its kerosene fuel. A rocket, on the other hand, carries its own oxidizer and is thus independent of the atmosphere. Nowadays we are used to hearing a range of devices, from fireworks to guided missiles and spacecraft, being referred to as rockets.

Early origins – rockets at war

The first rockets were probably made in China. Perhaps they were inspired by the 'ground rat', a type of firework that shot along the ground, belching flames from behind. This technology may have been applied to the flaming arrows sometimes used in Chinese warfare, so that it was no longer necessary to fire them from a bow. It was probably around 1150 that someone put the two ideas together.

By the thirteenth century, knowledge of gunpowder had entered the Islamic world (again, presumably from China), and the Mongols used gunpowder at the Battle of Legnica in 1241, marking an early appearance of gunpowder in Europe. Soon Roger Bacon would write down a formula for gunpowder, and from then on guns would play a major part in many battles and sieges.

Other writers continued to explain the idea of rocket propulsion. By the sixteenth century many of the key concepts of rocket design – the step rocket, the cluster rocket, and the rocket with wings – had been published.

The key development, making rockets more powerful, came in the eighteenth century with the work of the Indian prince Hyder Ali and his son Tippu Sultan. Hyder Ali started to build rockets with a metal container in which the fuel was burned. The greater strength meant that bigger explosions could take place, resulting in greater thrust. Tippu Sultan used such rockets against the British in battles in 1792 and 1799, alerting the West to a potent new weapon in the process. As a result, the British took up the rocket, and inventor William Congreve developed it.

Caricature of a nineteenth-century rocket

Looking into space

With improvements in artillery in the mid-nineteenth century, the rocket went into decline as a weapon. But at the end of the century a few visionaries could see another area in which the rocket had potential, and Konstantin Edouardovich Tsiolovsky started to write about the idea of space travel in 1895. He realized that the best way to create the power needed to achieve escape velocity (that is, sufficient velocity to 'escape' Earth's gravitational pull) was to use propellants made of liquefied hydrogen and oxygen.

Another important pioneer, the American R. H. Goddard, speculated interestingly about the potential of the rocket. In between making modifications to rocket engines, he imagined the day when a rocket would take a camera out into space to photograph the dark side of the Moon.

Goddard tested his first liquid-propelled rocket in 1926. It was an important step forward and enthusiasts around the world were fired by his work. During the next six years many societies for rocket research and the development of space exploration were formed. A member of one of these societies was the German scientist Wernher von Braun. He was making rockets capable of one-mile-high flights by 1931. Soon the German government seized on his work and he was developing rockets for warfare.

The team of scientists working for the German Army produced some formidable weapons, most notably the V2 missile. This weighed some 13.5 tons, had a one-ton warhead, and could cover a range of 200 miles (320 kilometres). It was first launched in 1944, and by the end of the war about 4,000 had been dropped.

After the end of the war it was clear that Germany held the lead in rocket technology, and both the USA and USSR wanted to capitalize on this work. Von Braun and a large team went to America to continue their work. Other German scientists went to Russia. The result was the independent development of rockets for the two countries' space-exploration and weapons programs, which would send satellites, vehicles, and ultimately people into space.

The rocket has changed the course of many wars – rocket-powered missiles are still an important weapon in the modern arsenal. They opened up a new chapter in human exploration. And they allowed the launching of satellites, which have transformed global communications. Their legacy is powerful and all-pervasive.

Stages in the development of the rocket

At the bottom of the picture are an early drawing of a space ship by Tsiolovsky and Robert Goddard's first liquid-propelled rocket. Above Goddard's head is a German V2 rocket. To the left are the multiple stages of a modern rocket and to the far right is a giant Saturn V rocket, designed to launch manned craft on their way to the Moon.

Multi-stage rocket
Ancestors of the rocket

NUCLEAR POWER

Scientists little guessed that splitting the atom would lead to an energy source of awesome potential

Scientists are used to the idea that chemical compounds are made up of molecules, each molecule being, in turn, made up of a combination of atoms of different elements. For example, each water molecule consists of two atoms of hydrogen and one atom of oxygen, hence its chemical symbol, H_2O.

Since the time of the Ancient Greeks, it was thought that the atom was the smallest particle, the basic building block of matter. Until the last century, scientists thought of it as a solid ball of matter. But during the nineteenth century scientists began to realize that the atom itself was made up of various subatomic particles. They started to look upon it as a sort of minute solar system, with a central nucleus of heavy protons (positively charged particles) and heavy neutrons (uncharged particles) orbited by negatively charged light particles, or electrons.

There have been various different 'maps' of the atom since the physicist Ernest (later Lord) Rutherford inferred the existence of the nucleus in 1906. But the important thing from the point of view of power generation is the idea that even the nucleus can be divided in two. When the nucleus of a heavy atom, such as uranium, splits into two or more parts in the process called fission, an enormous amount of energy is let off as a result of the conversion of mass to energy, in accordance with Einstein's formula $E = mc^2$. This equation is the basic principle of nuclear power.

'Splitting' the atom

Physicists such as Rutherford (in 1919) and Sir John Cockcroft (in 1932) modified nuclei by bombarding them with protons. But they could not imagine their experiments leading to a harnessing of energy produced by the process – on the contrary, they had to use more energy to divide the nuclei than was ultimately given off by this division.

Rutherford's apparatus for 'splitting' the nuclei of nitrogen atoms

The steps that were to take an important academic discovery straight into the realm of everyday practicality were made by a group of scientists working on the heavy element uranium. Naturally occurring uranium exists in three different isotopes. (This term is used to indicate atoms of the same chemical element that have different numbers of neutrons.) The physicist Otto Hahn, his colleague Lise Meitner, and her nephew Otto Frisch worked with the uranium isotope known as uranium-235. This is an unstable isotope, and the only uranium isotope that will undergo fission. Hahn discovered that by bombarding it with neutrons, you cause it to break down into two equal parts (called fission fragments), together with two or three neutrons. These extra neutrons encourage the disintegration of further uranium atoms, creating a chain reaction. In addition, energy is released by the breakdown. The question was further investigated by two physicists who emigrated to the USA from Nazi Germany in the 1930s, Leo Szillard and Enrico Fermi. Building on the work of Frisch and others, they set about creating a controlled, continuous source of nuclear energy, called a pile, which they did for the first time in 1942.

The lasting influence

These early reactions produced another element, plutonium, as a by-product. This too is a fissile element, and the two elements, uranium and plutonium, would be used in nuclear weapons. Since the research of Szillard and Fermi came to fruition in war-time, it was this aspect of their work that was seized upon by the US government.

But the splitting of the atom had more positive effects after the Second World War. By 1956 the first, practical, nuclear power station was working at Calder Hall, England. It and its successors use the energy produced in the nuclear action to produce steam, which is in turn harnessed to drive turbines and generate electricity.

With such a potentially hazardous fuel, and a process that releases vast amounts of energy and creates dangerous plutonium as a by-product, safety has been a key challenge – indeed, many would say a key problem – with nuclear reactors. This has been the case particularly in places such as some of the formerly Communist, eastern European countries, where monitoring and design have been so poor, and money so hard to come by that proper safety procedures have not been followed. But even in the more affluent, safety-conscious West there has been concern, and nuclear reactors are not built as often as they were in the 1960s and 1970s. Nevertheless, they have provided a significant proportion of the world's energy requirements. And as the problems of traditional power stations (diminishing fossil fuels and multiplying amounts of greenhouse gases) continue, they seem destined to continue to play their part in supplying the world's energy needs for some years to come.

Antoine Becquerel (1896), while studying X-rays, found that uranium emits invisible radiation.

Max Planck originated the quantum theory (1900), that energy exists in small exact units called quanta.

Niels Bohr (1913) described the atom as a nucleus with orbiting electrons.

Ernest Rutherford (1919) deduced the presence of the nucleus of the atom.

James Chadwick (1932) discovered neutrons, electrical neutral particles in the nuclei of atoms.

Otto Hahn (1939) produced the first nuclear fission without realizing the implications of the reaction.

Lise Meitner (1939) realized the implications of her own and Hahn's work on fission, and split atoms so that energy was released.

Enrico Fermi (1942) created the first atomic pile or nuclear reactor.

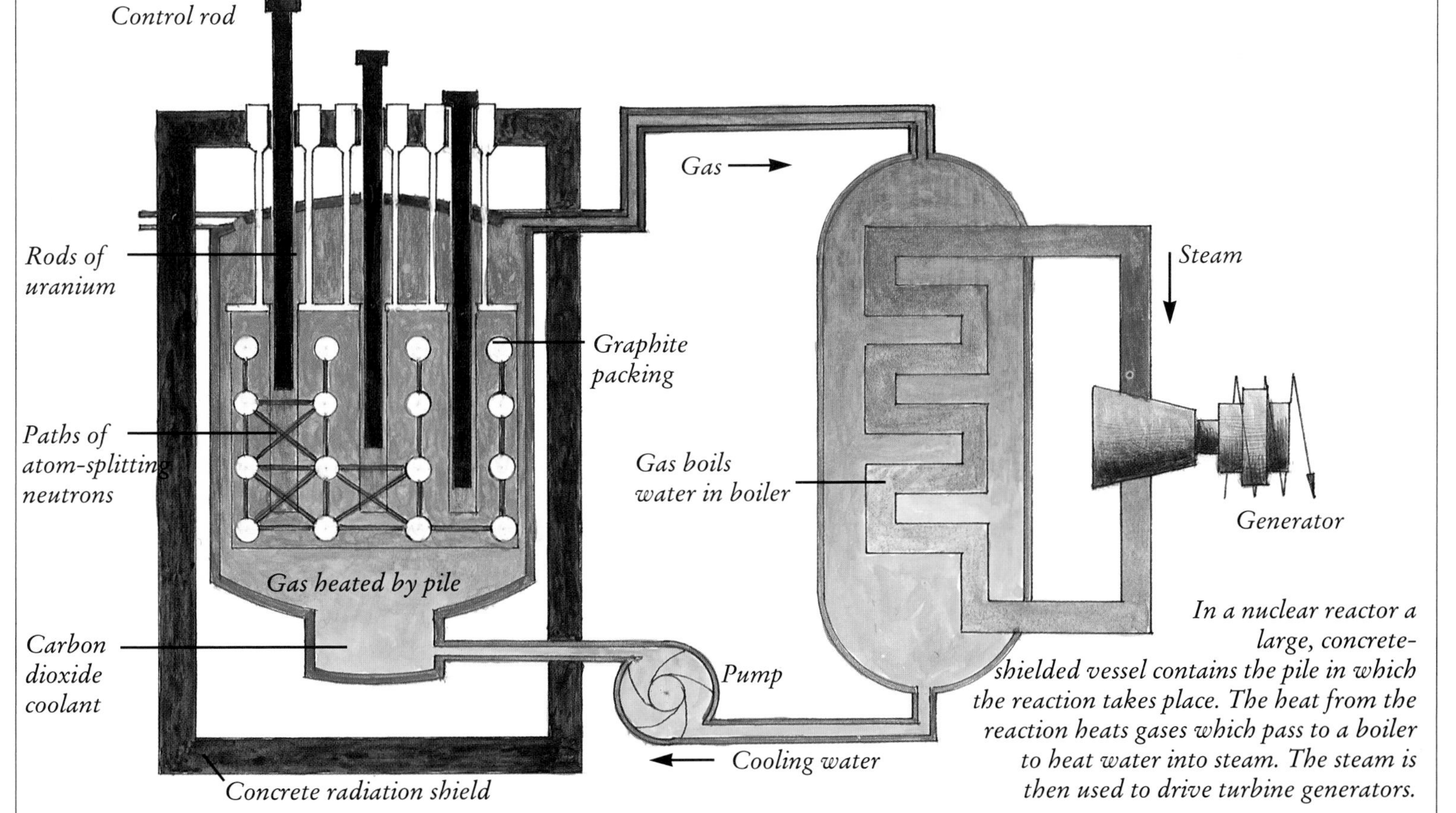

In a nuclear reactor a large, concrete-shielded vessel contains the pile in which the reaction takes place. The heat from the reaction heats gases which pass to a boiler to heat water into steam. The steam is then used to drive turbine generators.

USING NATURAL AND ARTIFICIAL MATERIALS

Wood, stone, leather: natural materials have provided some of our most useful resources. From wood can be made a seemingly limitless range of objects, from buildings to toothpicks; it has also been used widely as a fuel. Stone has been a building material for thousands of years, and has been used to make smaller objects, like vessels. Animal skins have long provided clothing; in earlier societies they were used to make tents for shelter, or parchment for writing.

Such materials could be shaped with little preparation. But there are others, equally natural, that need to be transformed into some new, artificial substance. An important example are the fibres that can be spun and woven to make cloth. Finds of stone spindle whorls and loom weights tell us that the skills of spinning and weaving evolved early – during the Neolithic. The oldest surviving piece of cloth is woollen material from the settlement at Çatal Hüyük in Turkey, and is some 8,500 years old. Cotton was being grown in the Indus Valley by around 2500 BC.

Many modern textiles are made in the same way as these early examples. The process of transformation from fibre to thread to woven cloth is much the same. Another influential early transformation was carried out by the potter. This was a different sort of process. For the textile worker, the challenge was a mechanical one: how to rearrange the fibres into a cloth of the required thickness and size. For the potter the discovery was physical: clay gets hard when it is heated. Once this discovery had been made, vessels for practically every purpose could be produced in this way. Their main disadvantage – they were easily broken – was offset by the long-lasting nature of the material, its ease of production, and its wide availability.

Heat also helped in metalworking. The first metalworkers hammered items out of small pieces of metal ore taken straight from the ground. But it was discovered that heat helped one to obtain the pure metal (by smelting), and that it also helped in producing metal objects (by casting). As a result of these discoveries, metalworking became an industry that could supply a huge range of human needs.

The discovery of these physical transformations probably came about by accident, and may well have happened at different places and times independently. There was no search to fulfil a specific need – people managed quite happily with wooden and stoneware vessels and stone tools until the potters and metalworkers came along.

The stories of the modern artificial materials, like plastics and artificial fibres, also include happy accidents. But they also reveal scientists worrying at the theory of their substances, designing molecules and then working out how to make them, as well as benefiting from the chances of the workshop. In whatever way the discoveries are made, they have transformed our lives, providing materials for items as diverse as cars and computers, ropes and clothing.

POTTERY AND GLASS

The invention of non-porous materials was an important step forward in the preparation and storage of food

Early peoples often led a nomadic life, looking for food to hunt and gather, and relying on temporary shelter from the elements and enemies. For such people, vessels and containers to carry food and drink were obviously important. They could make bottles out of skins, or weave baskets to carry solid food. Both types of container were light and strong and thus well adapted to life on the move.

But around the same time that people began to settle down, cultivate the land, and build themselves more permanent homes, they began to make their containers out of a different material – clay, which, when heated, loses its chemically bound water to form a hard, porous substance suitable for pots. Clay pots are relatively fragile and heavy – they are not suited to life on the move. But for a sedentary people they have clear advantages for protecting food from the weather and vermin. They can be made easily, the raw material is readily available in most places, and the pots can be made in a wide range of shapes and sizes. Pottery was therefore one of mankind's most important inventions.

The beginnings

The oldest pottery so far discovered is from Japan, dating to around 10,000 to 10,500 BC. These fragments, found in Fukui Cave, Kyushu, already look quite sophisticated. They are decorated with marks made with the fingertips or the fingernails, and with lines scraped in relief. The tradition that produced this pottery was probably well established, and there are likely to be still more ancient fragments waiting to be found, either in Japan, or on the mainland of Asia.

Pottery was certainly being produced in China and Indo-China soon after these Japanese examples; elsewhere it came later. In Africa and the Middle East the potters were at work by the eighth millennium BC, in Europe by the seventh. We do not know if pottery was invented independently in these regions, or whether it spread gradually westwards, perhaps along the coasts, aided by sea travellers. America's first pottery was produced in the fourth millennium BC, and was certainly an independent invention.

Early Chinese vessel

The early methods

There were various ways of making a vessel out of clay with a minimum of equipment. One of the simplest, and probably the earliest was to take a lump of clay, make a hole in the centre, and pull up the sides to form a pot. Later, the technique of coiling was developed. Strips of moist clay are rolled into shape and then laid on top of each other in circles, or spirals, to produce the body. This can be built up on a base made of a solid piece of clay, or the base itself can be built up as a spiral of strips. The sides of the piece can be smoothed down as the pot grows, so that the strips are joined together. This conceals the joins between the clay strips, and gives a flat surface for decoration.

Early potters often made flat dishes using some sort of mould. This could be a simple stone held inside the pot, against which the clay could be hammered when half-dry, rather like working metal on an anvil. Or it could be a properly shaped mould (perhaps a basket, or an already existing piece of pottery) which would give the pot its precise shape when the clay was pressed against it.

Whichever method was used to make the shape, the pot would then have to be hardened. To begin with, pots were probably simply left out in the sun to dry; this meant potting had probably to be done in the summer months in northern latitudes. Alternatively, pots could be put in the embers of a wood fire to harden.

From moulding the clay to firing, then, it was possible to produce these simple vessels with very little equipment. Once the techniques were well known it was possible for anyone to make and fire their own pots, but potting became an art quite early on, often practised by the women of the family or tribe. From the more successful potting families, the future specialist potters would have descended.

The potter's wheel

Building up a pot by coiling was easier if you could keep turning it around. Potters realized that a rotating platform on which to stand the pot would simplify the job and make the pot more symmetrical. The first basic potter's wheels, often called tournettes, were simply this, rotating platforms moving on a short pivot that was fixed to the ground. They were not designed to move fast, and they did not radically change the way pots were made.

The next stage was to make the wheel higher, and to devise some way of making it stable, so that the pot would be at a more comfortable working height. The result was something that looked more like the

The appearance of the wheel liberated the potter, making possible the creation of all sorts of rounded vessels. Such simple potter's wheels are still used in many places today.

potter's wheel of today, but which was still essentially a rotating work-bench.

Then someone realized that putting a disc on the bottom of the turning shaft would enable the potter's feet to turn the wheel. Suddenly the wheel would be able to turn much faster, while the potter's hands were still free to work the clay. A fast-spinning wheel like this has an additional advantage. The rotation creates a centrifugal force, which can be balanced by the pressure of the potter's fingers on the clay. This makes shaping much quicker – and easier – once the skill of 'throwing' the pot has been mastered.

The tournette appeared in Mesopotamia around 3500 BC. The potter's wheel proper came much later, in around 1500 BC. The latter was the more significant invention, and played an important role in turning pottery into a specialized craft.

Firing the clay

But there was still one more important step in the evolution of the craft of pottery – the development of the kiln. Pots dried in the sun, or with the use of an open fire, can be perfectly serviceable – but they are not perfectly watertight. To make permanently watertight pots you need a kiln to create higher firing temperatures. At these temperatures the pores in the clay fill up with molten paste, which sets to form a watertight barrier.

The kiln developed first in Asia. In China, for example, kilns were being used in around 4500 BC. In Mesopotamia, kilns around 6,000 years old have been found; similar ones were being used in Egypt about 1,000 years later.

These were all vertical kilns, consisting of a lower chamber that contained the fire, connected by a vertical flue to an upper kiln chamber where the pots were stacked for firing. The top of the kiln chamber would be open, to allow the potter to put in the items to be fired. When they were all in place the top could be sealed with pottery fragments from previous firings, to keep the hot air inside the kiln. The provision of a fixed roof on top of the firing chamber would allow still hotter temperatures to be reached and maintained.

Kilns like this could be dug out of the ground, or built of clay. Either way, they are often well preserved, since repeated firings have effectively made them into large pieces of pottery themselves.

Just as using a fast potter's wheel required particular skills, so did using a kiln. It was not simply a matter of putting the pots in and lighting the fire. To avoid underfiring or overfiring, the shape of the kiln and the position of the flue and the fire probably had to be modified to get the best distribution of heat. An experienced potter would find that particular effects could be obtained by putting a pot in a certain part of the kiln, or by loading the kiln in a particular way.

The use of glazes

Potters had other ways of making pots less porous. One was to use a 'slip', in other words to immerse the dried pot in a smooth solution of clay before firing. But a later and more widespread method was to use a glaze. This was a substance that was applied to the pot before firing (or between two firings). Among the most popular glazes were copper and lead, and they had the added advantage of making the pot look more attractive.

A tablet found at Tell'Umar in Iraq and dating to *c.* 1600 BC contains a formula for an early glaze. Copper was one of the earliest

Porcelain

The hard, white, glass-like substance we know as porcelain, though very different from earthenware, developed out of the same manufacturing process. It was first made in China, where kaolin, or china clay, was first discovered, and where the potters had kilns capable of firing it to very high temperatures.

What happens when kaolin is fired at a temperature of 2,336°F (1,280°C) or more is a process called vitrification. When this happens the clay changes its composition to produce the characteristic hard, translucent substance that today we think of as 'china'.

Chinese potters learnt that they could improve the process by mixing the china clay with a feldspathic rock called petuntse, which could also be used for glazing. Petuntse contributes to the process in two ways. It contains an alkali that lowers the temperature needed to start the vitrification, and it contains extra silica, which helps to make the porcelain more translucent.

The Chinese probably first made porcelain in the first century AD. It was one of the marvels of the East on which Western travellers reported back when they visited China. People who knew glass, but could not conceive of opaque clay producing such a glass-like substance were amazed. Chinese porcelain vases and bowls became some of the most prized objects to travel out of the East. But the Europeans were unsuccessful at finding out from the Chinese how to duplicate their results. Western potters tried to make porcelain, some believing that if they could find white clay they would succeed. But porcelain was not made in the West until the eighteenth century.

substances used for glazes, but later lead, tin, and salt were also used effectively on earthenware.

The techniques described here – throwing, firing, and glazing – all combined to make pottery a highly skilled and increasingly specialized craft. It is not surprising, then, that it developed in tandem with the world's great civilizations, where specialization was a way of life. In addition, pottery, because it is such a long-lasting material, has helped tell archaeologists much of what they know about the cultures that produced it.

Glass

Another ceramic substance, glass, was invented much later than pottery. This is not surprising; glass is a man-made substance comprising several ingredients – usually silica (sand), sodium carbonate and calcium carbonate (limestone) – which have to be heated together. A usable version of the formula was probably first made up in the Middle East, perhaps around 2600 BC, from when a glass rod found at Eshnunna in Babylonia is thought to date. Certainly the earliest fragments of glass yet found come from Mesopotamia, Syria, and Egypt.

Glass was not easy to make in the early days. Furnaces were small and hardly hot enough to melt the ingredients. It was hard work and some of the results were crude, although many were lovely. It was the Egyptians who seem to have been the first artists in glass, and some beautiful complete vessels, often luxurious cosmetic containers that belonged to members of the royal family, have been preserved. There was a glass workshop at the palace of Pharaoh Akhenaten, who reigned in the fourteenth century BC.

The Egyptians had an ingenious method of working their glass to the required shape. A clay shape was made and attached to a metal rod held by the glassmaker. Then the glass was built up on this shape, first in one colour, then in another layer of different-coloured threads. These threads were then pulled out of line with a comb-like tool, to give wavy or zigzag patterns.

The Mesopotamians used a different technique. They would sometimes cut a whole dish or vessel from a solid lump of glass, as if they were carving stone. There are some fine extant pieces from the eighth century BC made using in this technique, but the tradition probably goes back much further.

These methods of glassworking were passed through the Aegean civilizations and Alexandria (where glass vessels were made using moulds) to Ancient Rome, and the most prolific glass industry of the ancient world was based in the Roman Empire. The Romans seized on perhaps the most important development in the field since the invention of glass itself – the technique of glass-blowing.

Glass-blowing
The traditional method of making glass vessels was to blow the glass. Everyday items like wine glasses were made by blowing the shape of the bowl in a mould. A foot would then be shaped and added, and the surplus glass trimmed from the top edge of the bowl.

Glass-blowing was probably invented around 30 BC. It is not known exactly where glass was first blown, but it may well have been in Syria, where much early glass was made. Nor is it known how the idea first appeared. Glassworkers were used to holding the glass they were shaping on the end of a metal rod, and presumably these rods were sometimes hollow. Someone must have realized that blowing down the rod forced an air bubble into the glass, shaping it without the need of a solid mould. What was more, a blown bubble of glass could itself be shaped, meaning that fittings like handles could be made as easily as vessels.

Wherever the first glass-blowers worked they had a ready market for their wares – the Roman Empire. Soon, glass vessels were being produced in enormous numbers. They ranged from simple dishes and jars to elaborate decorative vases and cosmetic containers for the rich. Nearly all these wares were strikingly coloured, with the iridescent finish produced by impurities in the materials that makes early glass so beautiful.

But Roman glass, and the glass produced by the Islamic workers who took over the technology after the seventh century AD, was not the perfectly smooth, transparent material that most glass is today. And it could be made in only comparatively small pieces. As a result, some of the uses to which glass is now put had to wait for centuries. Glass for windows, for example, caught on only gradually, and was the preserve of the rich.

Even in the Middle Ages, glass windows, made of many small sections joined together with strips of lead (like the stained-glass windows still seen in churches) were among the most valuable items in the homes of the aristocracy.

Egyptian vessel of sand-core glass, sixteenth century BC

Roman flasks used for cosmetics

Renaissance glassware for the distillation process

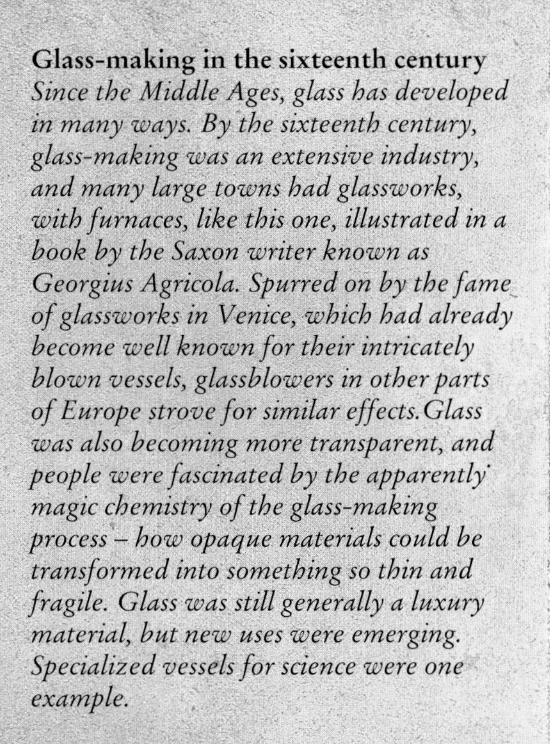

Glass-making in the sixteenth century
Since the Middle Ages, glass has developed in many ways. By the sixteenth century, glass-making was an extensive industry, and many large towns had glassworks, with furnaces, like this one, illustrated in a book by the Saxon writer known as Georgius Agricola. Spurred on by the fame of glassworks in Venice, which had already become well known for their intricately blown vessels, glassblowers in other parts of Europe strove for similar effects. Glass was also becoming more transparent, and people were fascinated by the apparently magic chemistry of the glass-making process – how opaque materials could be transformed into something so thin and fragile. Glass was still generally a luxury material, but new uses were emerging. Specialized vessels for science were one example.

Bottle-making
Bottles were originally made by forming the vessel on a core of sand. Then people learned how to blow vessels by hand. With the Industrial Revolution came automatic bottle blowing, in which the shape was gradually blown into a pre-formed mould.

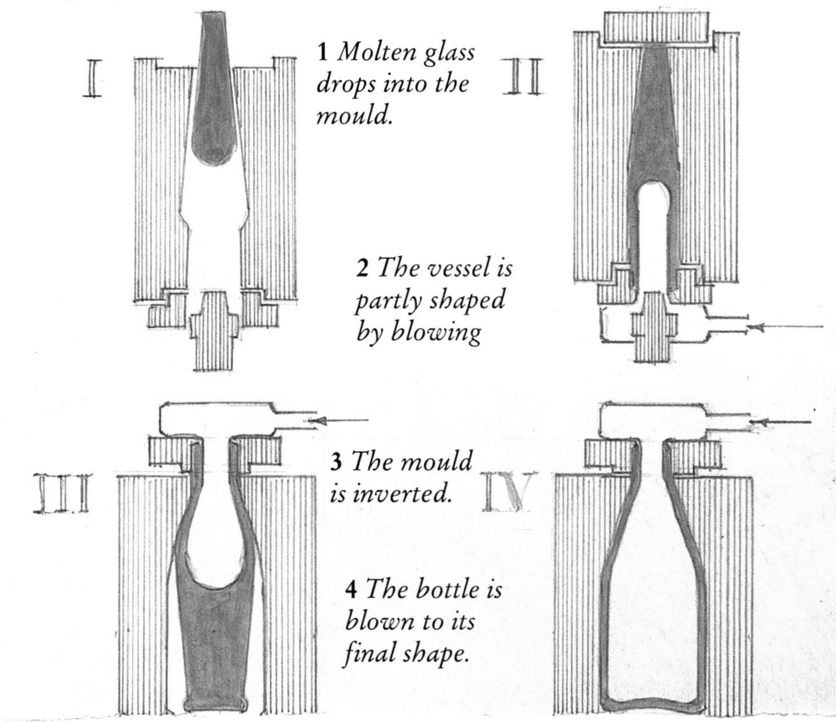

1 *Molten glass drops into the mould.*

2 *The vessel is partly shaped by blowing*

3 *The mould is inverted.*

4 *The bottle is blown to its final shape.*

Sheets of glass
Early glassmakers found it impossible to produce glass in really large, flat sheets. One early method was to pull a strip of molten glass from a furnace up the side of a tower. This produced glass that was good enough for house windows, albeit with the distortions typical in original eighteenth- or nineteenth-century glass. Later glassmakers used rollers to make a continuous ribbon of glass. But even this produced glass with roller marks that had to be polished away to give a perfect finish. The answer came as recently as the 1950s, with the concept of float glass. This involves floating the glass on the perfectly flat surface of a liquid (usually molten tin), which gives perfectly flat glass. The process was invented by British scientist Alistair Pilkington and is now used wherever a flat glazed surface is needed.

Heaters
Cooling
Annealing
Molten glass

WORKING IN METAL

Human life was transformed by the discovery of malleable materials that could be used to make practical and decorative articles

Making a clay pot by taking the raw material out of the ground, fashioning it into the required shape and heating it to make it harden was a step of major importance. Early craftworkers would soon have learnt which was the best type of clay to use, and this would have involved trying all sorts of samples from the ground. Before long, someone must have come across metal a curious, hard substance, at once different from both the stone used to make tools, and the clay that was best for pots. When a large enough deposit was found, it was realized that it was a malleable material and could be hammered into shape to make different objects. The first items to be made were probably small items, such as pins, and examples have been found in the Middle East (particularly in Turkey) that date from before 6000 BC.

In itself, this was a great advance. A new group of materials had been discovered that would transform human life. Before long, it was clear that there were distinct advantages in using metal rather than stone to produce tools. A variety of artefacts could be made with stone, but their shape depended on the way the stone split when it was worked. Metal did not have this limitation and so more functional shapes could be made. One example is saw teeth, which were always rather coarse when made of stone, and much finer in metal.

The first metals to be discovered and worked were copper, silver and gold. None of these was plentiful, but all were easy to work. Copper was the most widely available of the three. It was discovered independently in different parts of the world – in the Old World in Anatolia, and in the New World by native North Americans who found and worked the extensive deposits of copper on the southern shores of Lake Superior. In some places in North America iron was treated in the same way.

Bronze bull's-head, C. Asia, c.2250 BC

From metalworking to metallurgy

These were vital advances, but the true imaginative leap was made when people grasped the concept of an ore – a rock containing metallic deposits – which could be smelted to produce usable metal.

Again, this discovery was made independently in different places and at different times. In western Asia and south-east Europe it was probably around 6000 BC. In eastern Asia it came before 2000 BC. Knowledge of ores and smelting came to other areas of Europe, Asia and northern Africa from these original centres.

The first metal to be smelted was probably copper, which is not a common substance. Neither is tin, the metal that is added to copper to make the alloy bronze, which is harder, less subject to corrosion, and has a lower melting point that makes casting easier. So the development of metallurgy went hand-in-hand with the development of trade, which in turn was tied up with the growth of cities and civilization. Bronze was first used in western Asia in the fourth millennium BC, and knowledge of bronze spread along the trade routes of the Old World during the following millennium. It may also have been produced independently in eastern Asia slightly later than in the West.

The arrival of smelting had another advantage. When the ore was heated and the molten metal flowed out, the latter could be poured into moulds shaped like the objects required. Bronze was easier to cast than copper, and so the bronzesmiths heated their metal in a crucible over a strong fire and poured the molten alloy into moulds of clay or stone. In this way they could make solid objects like axe-heads or spearheads – items previously available only in stone.

Another casting process was invented to create hollow items such as bowls, which had previously been made by cold-hammering metal. This process has become known as the *cire perdue* (lost-wax) method. It involves moulding the shape of the object in wax and then setting the mould within a clay block. The wax is then melted and poured away, leaving a cavity of the right shape and size into which the molten metal can be poured.

These techniques of metalwork allowed some types of objects to be made that were difficult or well-nigh impossible to make in stone. For example, stone axe-heads and arrowheads had to be bound to their hafts with thin twine. Metal allowed socketed heads to be designed so that the haft could be inserted into the head, a more reliable way of keeping the two pieces together.

So, many items could be made in bronze, which became a vital resource for early civilizations. Because of its rarity, and its near-absence in the river valleys of the Middle East, where so many early cities sprang up, long trade routes were established to bring metal to places where it was in demand. The importance of these routes and the metal they brought,

Metal tools

Early metal tools display features that were unknown in their stone equivalents. Saws have fine, sharp teeth, knives have delicately made handles that fit the hand, and axe-heads have sockets to take their hafts.

together with the durability of bronze objects, led archaeologists of the past to talk about a 'Bronze Age'. But this is now seen to be a somewhat misleading term. Bronze came to different places at different times (to South America, for example, as late as *c.* AD1100, with the Incas), and for all its importance, it was always something of a luxury material. Many people carried on using tools and weapons of stone even when bronze was available. But a more widespread revolution was on the way with the exploitation of iron.

The coming of iron technology

Iron is much more common than copper or tin. It was probably first smelted in western Asia between 2000 and 1500 BC. There is a strong tradition that the Hittites of Anatolia were the first ironworkers, and that they guarded the secret of this metal jealously so it could not be copied by their rivals. Today archaeologists reject this tradition. Iron objects dating from before the Hittite period have been found, and there is no evidence for the supposed secrecy about the technology.

What is certain is that knowledge of iron spread across the world between 1500 and 600 BC to Europe, northern Africa and central and southern Asia. In these areas bronze was still used, but it became even more of a luxury material, used mainly for ornamental purposes in jewellery and other status objects. Meanwhile, better tools and weapons with sharper edges were being made from iron.

The method used was different from that normally used with bronze. After smelting, iron was usually beaten into shape, and its strength increased with each hammering. The resulting 'wrought iron' was widely used in the West until the time of the Industrial Revolution.

Higher temperatures were required to cast iron, and consequently it was not until the Middle Ages that cast iron was produced in Europe. But the story was different in China, where it was made as early as the fourth century BC.

How could the Chinese cast iron so early? First, they knew how to make blast furnaces, and the clays available in China kept the heat in well, so they could achieve higher temperatures than could their Western counterparts. Second; they found a way of reducing the melting point of iron, by mixing in a substance called 'black earth', which contained phosphate. Adding phosphorus in sufficient quantities reduced the melting point from 2066°F (1130°C) to 1742°F (950°C).

So, Chinese ironmasters began to produce a range of tools, utensils and vessels from cast iron. There were many immediate advantages, from longer swords to cooking pots with thinner walls. There were also further items that had never been made in metal before, such as cast-iron ploughshares, which transformed other areas of Chinese technology.

Chinese ironworkers also addressed themselves early on to the main problem of cast iron: it is more brittle than wrought iron. As early as the third century BC they discovered the concept of annealing (keeping the metal at a high temperature for a period of up to a week), which made the iron more malleable and less brittle. It meant that an item like a cast-iron ploughshare was more practical because it would not shatter when it hit a large stone.

The Chinese used cast iron in some surprising ways. They made entire pagodas

The earliest known drawing of a blast furnace from China dates from 1334, but this illustration also takes information from later sources. Four men operate the two large, hinged bellows that supply the blast. The furnace itself was fed with iron ore, coke or charcoal, and limestone. As the molten metal emerged at the base, it ran into a large, square, stone trough. Here the liquid metal could be stirred, the first stage in making it a stronger, more malleable material. An alternative arrangement sent the molten iron from the furnace directly into moulds, where it could set to make pig iron.

out of it, casting each storey separately before joining them together on site. One empress even had a 105-foot- (32-metre-) high, cast-iron column built in AD 695. In AD 954 Chinese ironworkers made their structure with the largest single piece of cast iron, the Great Lion of Tsang-chou, which stands twenty feet (six metres) tall. So the Chinese were well aware of the vast potential of the material that they had mastered, using it in structures of a size that would be unthinkable in the West until the industrial age.

Iron in the west

The iron industry was rather more primitive in the West. The problem was getting the temperature in the furnace high enough to melt the iron. In the Greek and Roman world furnaces were fuelled with charcoal, and the heat was intensified by means of foot bellows. This arrangement did not produce enough heat to make the iron completely molten. Instead, a small ball of iron, called a bloom, formed at the bottom of the furnace. This could then be taken out of the furnace and hammered into a bar of wrought iron that could in turn be reheated and hammered again to produce whatever object was required.

Iron produced in this way was adequate, but neither as hard nor as strong as was necessary for the best quality knives and swords. So ironworkers started to use a different type of reheating process to improve the quality of the metal. They reheated the iron bars between layers of charcoal. This carburized the surface of the iron, in effect making a coat of steel (iron that contains up to 3.5 per cent carbon) around the bar. Still further heating and hammering of this 'case-hardened' material gave blades of high quality.

The Indians made this type of steel in a crucible. They placed their carburized iron bars in a closed crucible of clay and heated it, causing the carbon gradually to diffuse through the iron. This helped ensure a consistent carburization all the way through the bar. Such steel was highly prized by the Romans, and the process caught on in the West. But when the Roman Empire collapsed, the secret of steel production went with it.

By the Middle Ages, cast iron was more available in western Europe, as furnaces capable of higher temperatures were built. For better quality items (especially knives and swords) there were small quantities of case-hardened steel, but widespread availability had to wait until the Industrial Revolution in Europe. In the Western hemisphere, ironworking started with the arrival of the Europeans.

New and old techniques

One of the first and most significant steps in the European Industrial Revolution was the introduction of coke into the smelting process. Coke is coal that has been heated in a reducing atmosphere to drive out many of its impurities. Lord Dudley described the process in 1621, but it was not until almost a century later that it was applied to iron smelting.

This was the result of the work of Abraham Darby, who grew up as an apprentice to a maltster. Here he learned how coke was used in the malthouse, because the impurities in coal got into the malt and ruined the flavour. After learning the maltster's trade, Darby branched out. He started his own business making brass wire, but also began to experiment

Pattern welding

Although it was possible to cast an object by pouring molten iron into a mould of the required shape, cast iron was brittle, so this method was not ideal for many tools and weapons. An alternative was pattern welding, in which cast strips of iron were hammered together so the finished shape was built up gradually. Furthermore, the hammering gave additional strength to the iron.

Three iron rods are twisted together and the process of hammering begins.

Further hammering starts to give the object its shape, until finally a strong sword is produced.

with casting, taking out a patent for casting pots in sand in 1707. Two years later he set up a foundry in Coalbrookdale in Shropshire, a depressed area where there were already iron foundries losing business because of the shortage of coal.

Darby's furnace was fired with coke; he had realized that there were several advantages to using this fuel. For a start, the local coal was suitable for coke-making. Coke was also strong, so a large quantity of ore could be put into the furnace without the fire collapsing. This in turn allowed larger objects to be cast. Darby's business was an immediate success.

The next key step was around 1740, when Benjamin Huntsman heated a mixture of wrought iron and a precisely measured amount of charcoal in a crucible, thus rediscovering the crucible-steel process used by the Indians centuries before. With plenty of fuel now available to fire furnaces, the way was open to steel production on a large scale. As well as familiar pots and pans, knives and needles, steel could be used in the production of the machines that would further the Industrial Revolution, from steam engines to railway locomotives. One of the major elements of the transformation of the Western world was in place.

A changing influence

The development of metalworking was one of the most important in the history of human inventiveness, and it was important in different ways at different stages in the story. To begin with it represented a fundamental departure in human technology. Taking a material that existed in one state – as an ore – and transforming it by smelting into a useful material was a process unlike anything that had gone before. To the people of the time it must have seemed like magic; from a twentieth-century viewpoint it looks more like the beginning of the science of metallurgy.

Metal also had a profound cultural significance, arriving with city life and trade, and providing better weapons of war and status objects for the new upper classes of the ancient empires. With the exploitation of iron, metal objects reached a still wider group of users. A Roman kitchen, for example, contained many of the metal knives, bowls and other utensils that we still use today.

In the industrial era things changed again. The new material heralded a new age. First iron, and then steel seemed to symbolize more than anything else – except maybe the steam engine – the coming of industry, the expansion of communications, the growth of cities, and the success of new technological leaders.

The Bessemer converter

In spite of the success of Huntsman's process for making crucible steel, inventors continued to look for ways of making steel more cheaply and efficiently. A notable example was another British inventor, Henry Bessemer. In the 1850s he developed a process in which air was blown through molten iron to burn out all the carbon in the metal. This carbon was then replaced with the precise amount needed to form steel. The process was carried out in a large container called a converter. The molten iron was poured into the converter and the air was forced in through holes in the side. After the 'blow', manganese was added. This combined with surplus oxygen left in the metal as a result of the blow: removing the oxygen in this way meant that the steel was less brittle than it would have been without the manganese. The converter was constructed so that it could be tipped on its side to empty out the molten steel.

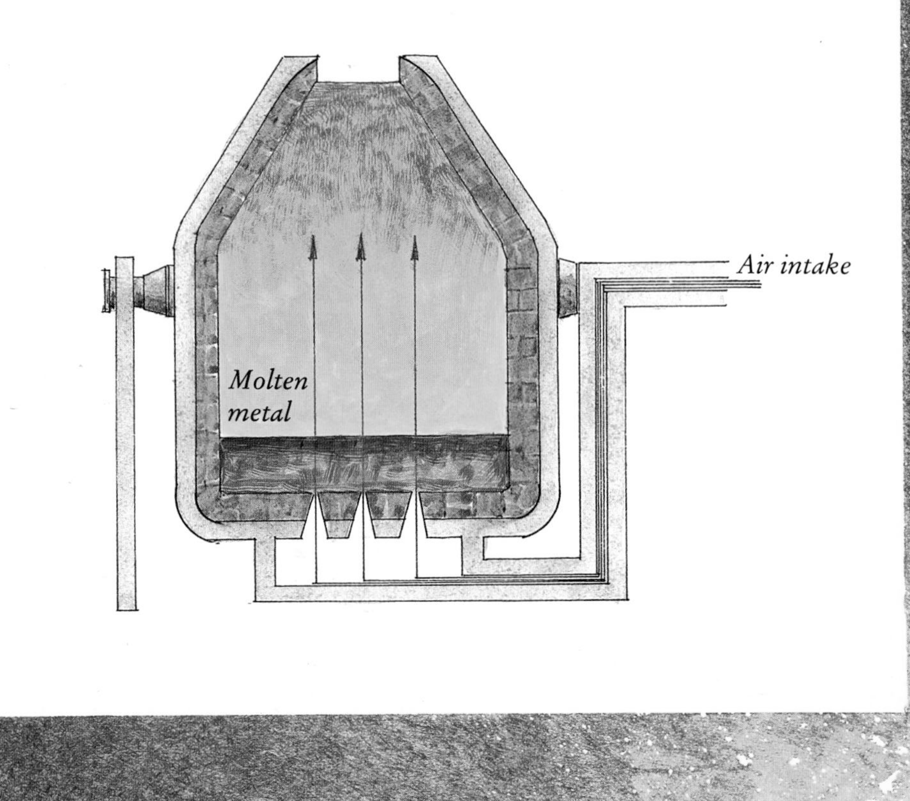

The open-hearth process

Pioneered by two German brothers, Wilhelm and Friedrich Siemens, and developed by Pierre and Emile Martin from France, this process was another influential method of steel production – and for melting down scrap steel for re-use. It incorporated preheaters, which heated up the air going into the furnace so that the overall temperature inside could be kept much higher.

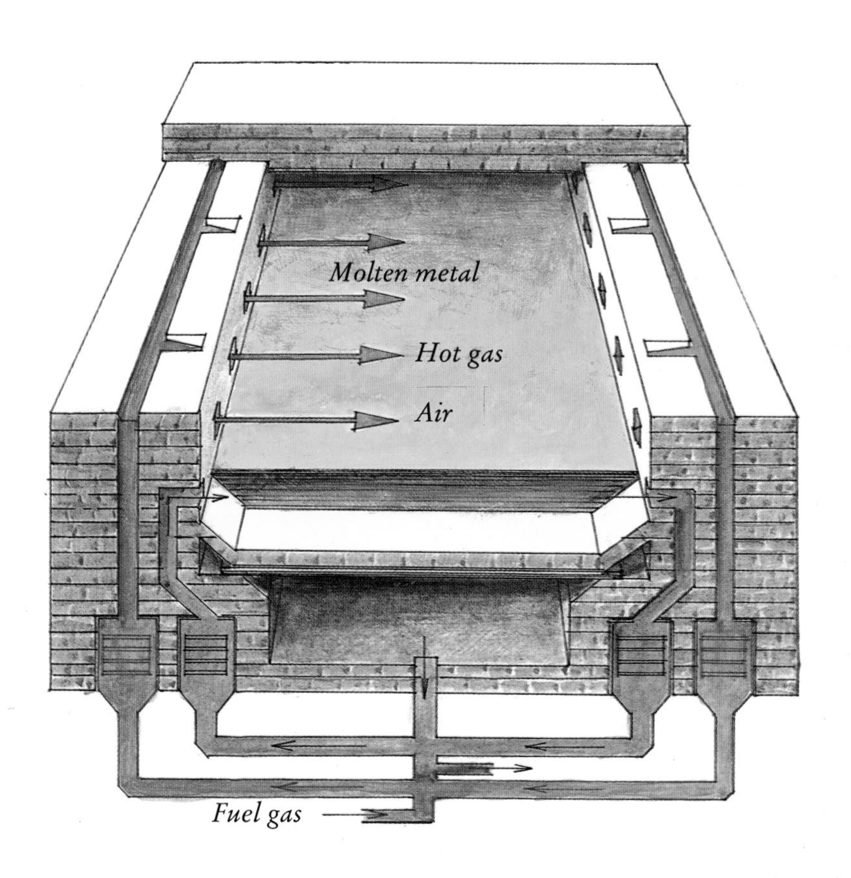

PLASTICS

From shirt collars to cinema film, few areas of life have remained untouched by plastics

Few revolutions in science have affected so many areas of life as did the growth of the plastics industry in the second half of the nineteenth century. These materials, which are made artificially, which can be bent or moulded under heat or pressure, but which subsequently remain stable, are now ubiquitous, being used in everything from packaging and clothing to toys and industrial components.

Rubber

The only similar natural product is rubber, not normally thought of as a plastic, but sharing many of plastic's properties. It is made of the sap from the tree *Hevea brasiliensis.* As this name suggests, it was originally a South American tree, but in the nineteenth century seedlings of the species were propagated, first at Kew in London, and then in Malaya. The latter area has remained the world's main rubber-producing region.

Natural rubber is a polymer, that is to say, it is made up of long, chain-like molecules, themselves made up of simple sections strung together. Its most obvious property is elasticity – it springs back to its original position after it has been stretched. This quality had been appreciated since at least the thirteenth century, but there was a problem: the natural material becomes sticky in hot weather and loses its elasticity when it is cold.

As a response to this limitation, American hardware dealer Charles Goodyear looked for ways to make rubber more workable. One experiment in 1839 involved a mixture of rubber, sulphur and white lead, which Goodyear one day overheated. The mixture charred but did not melt. Goodyear realized that if he could produce this substance in a uniform manner, he would have a material with great potential. And this he did by passing the material through a heated trough.

Goodyear did not find much enthusiasm for his material in the USA, and tried to interest Charles Macintosh, the British pioneer of rubber rainwear. But Macintosh's partner, Thomas Hancock, worked out Goodyear's process and patented the idea in Britain in 1844, before the American inventor could do so. Vulcanized rubber, as it came to be known, would soon be used widely as an insulating material, and for shock absorbers on the railways, as well as in its already well-established roles in clothing and shoes. The success of the material was enhanced in 1846, when another British scientist, Alexander Parkes, invented a different method of vulcanization. This proved suitable for making the sort of thin rubber that could be used in a range of other products such as balloons.

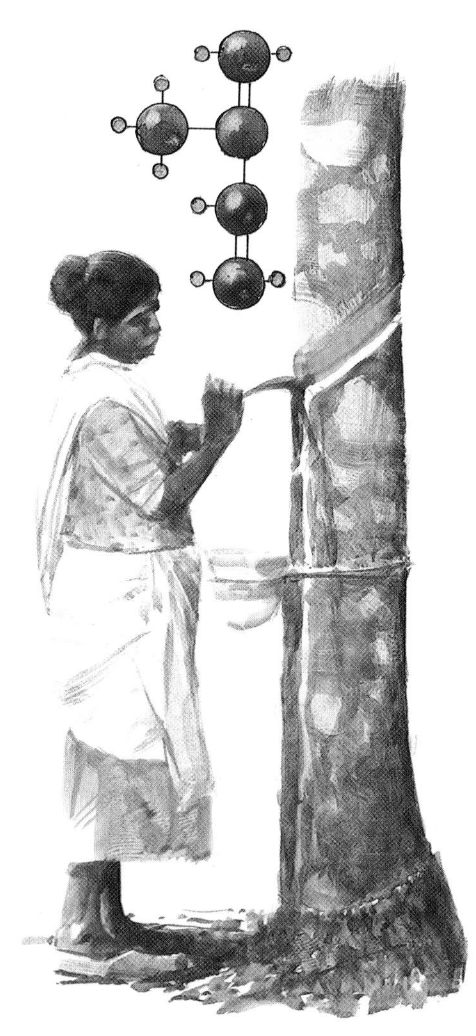

Extracting rubber

The early plastics

By the time of Goodyear's experiments, the groundwork that would lead to artificial plastics had already been done in Europe. Among the pioneers were the Frenchman Henri Bracconet, who, in 1832, poured concentrated nitric acid on to cotton or wood fibres to make a material he called xylodine. In Paris Théophile Pelouze created a similar material at around the same time. And in 1846 the Swiss chemist Christian Schönbein found a way of treating cellulose nitrate with nitric and sulphuric acids to produce 'plastic' properties.

At first, none of these discoveries met with much interest. Schönbein even sent some of his findings to Michael Faraday, writing, 'The matter is capable of being shaped into all sorts of things and forms and I have made from it a number of beautiful vessels'. But Faraday was not interested in Schönbein's work, and the task of exploring the properties of plastics was left to Alexander Parkes, who, as we have seen, was already working on the vulcanization of rubber.

Another of Parkes's interests was photography. This had led him to experiment with cellulose nitrate, making collodion (a solution of cellulose nitrate in alcohol and ether) to support a photographic emulsion. In the 1850s this work took Parkes further towards plastics. Mixing the cellulose nitrate with camphor, he produced a hard but flexible transparent material that became known as Parkesine.

By the end of the decade Parkes was able to produce a wide range of items with his new material. With the International Exhibition in London in 1862 he saw his commercial chance. He exhibited buttons, combs, pens, boxes, decorative pan-

Vulcanized rubber

Natural rubber is made up of a polymer of a substance called isoprene. Molecules of isoprene (1) are also used in the manufacture of synthetic rubber. In rubber these molecules are joined together to form long, chain-like molecules (2) that can slide over each other. This sliding gives rubber its ability to stretch easily.

In vulcanization the chains are joined together by atoms of sulphur (3). The chains, shown here as zigzag lines, can no longer slide over each other and the rubber is now hard and springy. Vulcanized rubber, which was pioneered by American inventor Charles Goodyear (below), is ideal for products, such as automobile tyres, where a tough, resilient quality is required. Electrical engineers soon seized on its insulating quality. It also became a valued material for items like shock absorbers on cars.

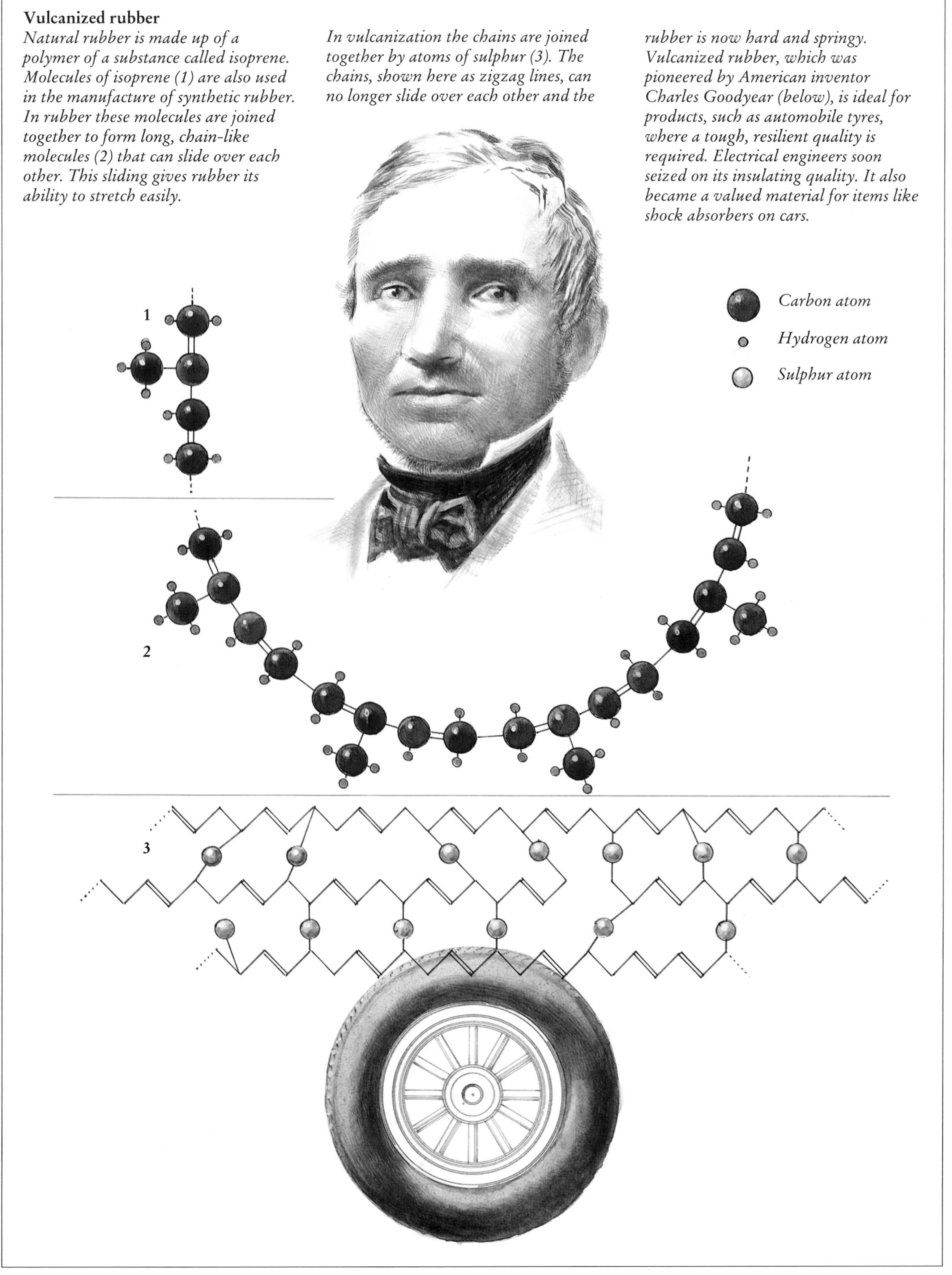

Alexander Parkes

The creator of 'Parkesine' was an inveterate inventor whose work embraced photography as well as rubber and plastics. In fact, the fields of photography and plastics were closely related in Parkes's work. The substance that Parkes came up with, variously known as Parkesine and celluloid, was nitro-cellulose plasticized with camphor. By 1888 it was being produced in sheets about one hundredth of an inch (0.5 millimetres) thick, which were ideal for use as a photographic base.

Cast phenolic plastic radio, produced in the USA in the mid-1930s.

els and even jewellery, all made of Parkesine. The accompanying advertisement extolled the versatility of the material, pointing out that it could be made hard or flexible, that it could be cast or used as a coating, that it was waterproof, and so on. It seemed a commercial certainty that the material would be a success, but Parkes's company went out of business – mainly because of poor quality control and Parkes's lack of business skills.

The business was taken over by his factory manager, Daniel Spill, but the next important development in the story of plastics was taking place on the other side of the Atlantic. In Albany, New York, a firm of billiard ball manufacturers were worried about the shortage of ivory. They put up a prize for anyone who could make acceptable billiard balls from an artificial substance. The response came from John Wesley Hyatt, a New York printer.

Hyatt met with some success. He worked out a more efficient manufacturing process and improved the material, which he sold under the new name of celluloid. Some of his improvements sparked off a dispute with Spill, who alleged that Hyatt was infringing some of his patents. Hyatt eventually won the battle and found a variety of uses for his plastic. One which was particularly successful was the manufacture of shirt collars and cuffs, which were valued because they could be washed and dried so easily. And so celluloid, as manufactured by Hyatt, remained the most successful plastic until the end of the nineteenth century.

According to a legend (which may be true), it was an accident that spurred on the next development. In the 1890s, so the story goes, a cat belonging to German scientist Adolf Spitteler knocked a bottle of formaldehyde solution into its saucer of milk. The result was a hard plastic. What had happened was that the formaldehyde had reacted with casein, the protein in skimmed milk. Spitteler and his collaborator Krische were soon producing this new plastic, which they called Galalith. It was water-resistant and could be made in a wide range of colours. Of all the early plastics, this was the one that heralded the widespread appearance of cheap products made of artificial substances that usually come to mind when we think of plastics today.

And so plastics began to come of age. At first they had been perceived as materials with which to make high-quality goods, and part of the reason for Parkes's business failure was that his products failed to live up to these expectations. But people soon realized that there were some things plastics could do well that other materials could not – whether it was producing easily washable shirt collars, or making insulators for telegraph companies. There were other uses for these early plastics, in fields that were themselves new. Photographic film was an obvious example, since there was a growing demand from the new cinema industry. Various different formulae were tried, including a non-flammable cellulose acetate film that was made by the Swiss brothers Henri and Camille Dreyfus as early as 1910.

Another new and fast-developing field where the new materials were useful was in the aircraft industry. Cellu-

Chemical structures
The diagrams show how molecules of the substance ethylene, with their atoms of hydrogen and carbon, can be polymerized to make different plastics with their characteristic chain-like molecules. In the first example, polyethylene is produced. In the second, the ethylene reacts with chlorine to produce polyvinyl chloride (PVC).

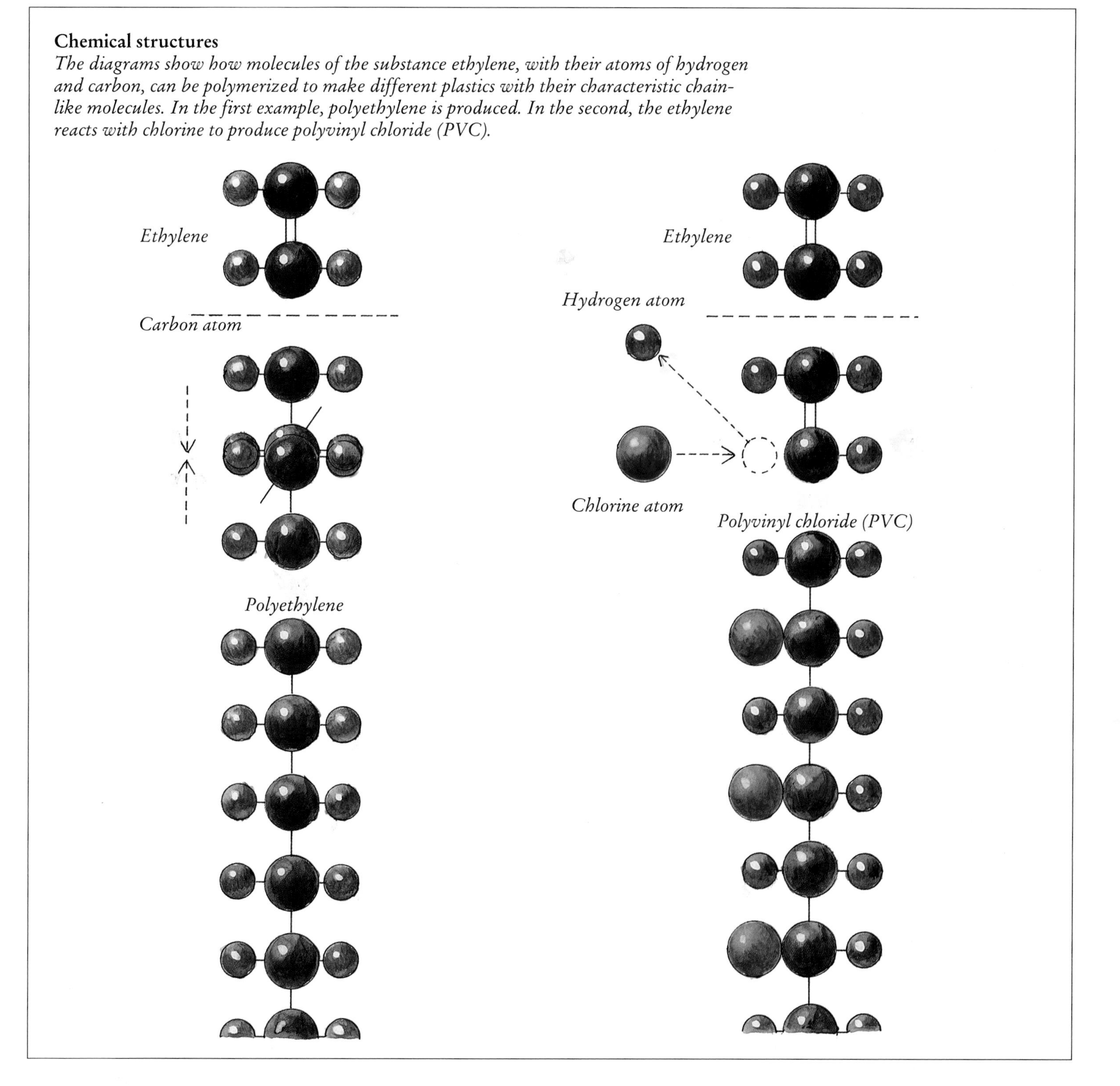

lose acetate was widely used as a 'dope' for aeroplane wing fabrics during the First World War. When the war ended, the resulting overcapacity led manufacturers to turn to the production of other artificial materials, in particular the first artificial fibre, rayon. Finally the potential of plastics as cheap materials was acknowledged.

Bakelite

Meanwhile, Leo Baekeland was at work. Baekeland was born in Belgium, the illegitimate son of poor parents. He managed to overcome the problems imposed by his background and gained his doctorate by the time he was twenty. But there was still the stigma of illegitimacy, and he felt that the only place that would truly accept him and his abilities was America.Baekeland started, like several other plastics pioneers, with an interest in photography. He created a photogrphic printing paper and sold the manufacturing rights in the material to Kodak. The large amount of money Baekeland made from this was invested in his scientific experiments. He was a true enthusiast for artificial materials, experimenting ceaselessly, and his work led to more than one hundred patents.

It was in the early years of the twentieth century that Baekeland did his experiments with formaldehyde and phenol, producing a synthetic shellac. The fact that these two substances formed a solid, resinous material had been noted before, but Baekeland was the first to control the reaction. He called the synthetic shellac product Bakelite.

The new shellac had a quality that had not previously been seen in plastics. It was a 'thermosetting' material. In other words, when it had been heated to a certain temperature it became very hard, and resistant to further changes produced by high temperatures. It was therefore possible to heat the substance until it softened, mould it to the required shape, and heat it further so that it set permanently. In addition, Bakelite was water-resistant, an electrical insulator, and could be readily cut or machined to different shapes.

As is clear from these properties, Bakelite was a highly versatile material. During the early decades of the twentieth century a host of Bakelite products was produced. Its hardness was well suited both to decorative items like clock-cases, and to machinery that had to take a lot of wear. Its property as an electrical insulator meant that it was widely used for plugs, sockets, and other electrical fittings. Its heat-resistance made it ideal for containers for hot liquids. Soon Bakelite versions of everything from food bowls to toilet seats could be bought.

New polymers

During the twentieth century research into plastics advanced at great speed. One reason for this was the work done in the petroleum industry, during which chemists discovered many of the chemicals that could be made from crude oil. They also began to analyse the chemical structure of some of the constituents of the early plastics, such as cellulose. This research led to the discovery and analysis of polymers.

One line of enquiry was into the process called 'polymerization', in which two or more molecules of the same compound unite to form larger molecules of a new compound with larger molecular weight. In the inter-war period two chemists in particular did crucial work in this field. The first was the German Hermann Staudinger, who polymerized styrene into the thermoplastic polystyrene. The second was American Wallace Carothers, who worked for the Du Pont Company.

Carothers was trying to find an artificial substitute for silk, a material that was increasingly hard to come by in the USA because of difficult political relations with the main supplier, Japan. Carothers came up with two ingredients, the soluble organic substance diaminohexane, and the crystalline solid adipic acid, which contain the four elements carbon, nitrogen, hydrogen and oxygen. When combined, they make a long molecule with a 'backbone' of carbon and nitrogen atoms, and with hydrogen and oxygen atoms arranged on either side. This material became known as nylon.

Much extra work was needed to turn Carothers' discovery into a material that could be made in bulk, and it was not until 1940 that nylon was made in commercial quantities. By this time it had been discovered that it could be made from readily

Heat-retaining Melamine plastic cup

available raw materials. Carbon and hydrogen could be extracted from hydrocarbons in petrol or coal, nitrogen from the atmosphere and oxygen from water.

The production of artificial fibres required processes quite unlike those previously used in the textile industry. Molten polymer was forced through tiny holes to create fine filaments of fibre. These were then cooled and stretched, a process that greatly increases their strength. The process made fibre that was light, strong and cheap to produce. What was more, the fibres that were made in this way did not need any elaborate further processing to make them into usable yarns, as many natural fibres do. They were 'instant' yarns, ready for use. Soon a whole new textile industry had been born from Carothers' work.

Development work in this area led to a number of other artificial fibres, known by trade names such as Terylene, Lycra and Orlon. Such fibres had an obvious impact on the textile industry, although this could be in surprising ways: nylon was hardly like silk, but it proved even more suitable for women's stockings than the natural fibre. But it was not just the textile industry that benefited. Artificial fibres have become invaluable in many areas, from the manufacture of ropes far stronger and lighter than had previously been imagined, to the making of very fine surgical thread. And in wartime its properties as a silk-substitute were remembered: thousands of parachutes have been made of nylon.

In many ways Hermann Staudinger made an even greater contribution to the development of plastics. He made a series of investigations into the structure of rubber and such plastics as polystyrene and polyformaldehyde, and his experiments eventually led to a deeper understanding of plastics than anyone had had before. Staudinger was awarded the Nobel Prize for chemistry in 1953 for his work in this field.

A world of new materials

Nylon and its related fibres were not the only materials to appear as a result of the work of these pioneers. Carothers, for example, wrote many articles on different polymers, particularly during the 1930s. One material he discovered was neoprene, an artificial rubber that was strong, durable, and relatively quick and easy to produce. It and similar products are still widely used where previously rubber would have been required.

Other synthetic rubbers followed. These included Thiokol, a product with a high resistance to solvents, which is used for flexible fuel lines (for example in the hoses on petrol pumps), and butyl, which is ideal for automobile tyre tubes because it has a very low permeability to gases.

Since the time of Carothers, research into plastics has continued. Many new materials have emerged, such as polyethylene (tough and resistant to chemicals and water) and acrylics (a range of materials from resins to fibres). They play roles undreamed of by the pioneers of the last century. For example, their biological inertness means that they can be used inside the body, for the provision of 'spare parts', and in a host of substances where resistance to insect attack is important. Transparent plastics are often used where glass would have been the only option in the past. And plastics are also used in such diverse applications as paints and car body parts.

An understanding of the chemistry of plastics has led to substances that can be tailor-made for new applications, so plastics will continue to be seen everywhere. Unfortunately, their advantages are offset by considerable disadvantages – they cannot be burned and are not usually biodegradable, so they present a serious waste-disposal problem.

Plastic products

The range of artificial materials that we know as plastics have entered practically every area of modern life. Indeed, there are many objects that most people in the Western world find it hard to imagine in any other material. Credit cards, photographic film and long-playing records were specifically conceived as plastic items, using the qualities of the material (its ability to be moulded into shape with the createst of ease, its flexibility, and so on) to great advantage. But there are also thousands of other items, from combs to kitchen utensils, toys to buttons, which, while available in other materials, are made most often in plastic.

AMERICAN
01/92 THRU

SHRINKING THE WORLD

Until quite recently, most people hardly travelled at all. They were born, lived, worked, and died in the same village; the other members of their family probably lived there too. The nearest town, perhaps only a few miles away, took many hours to reach on foot. Their world was tiny, but the whole world was vast and unknowable. For the few who did move around – military commanders pushing back the frontiers of empire, people with special skills, whose work took them from country to country – travel was not to be undertaken lightly. It was, as one writer put it, 'slow, expensive death'.

Travel was hard work even after the inventors of the ancient world had got to work on the problems. Take the wheel: it seems so obvious now that it is hard to imagine a world without it. Yet it was not always obvious. Civilizations like those of Ancient Egypt and Central America grasped the idea of the wheel, as we can see from toys that have come down to us, but they do not seem to have used it for transport.

This is less surprising than it seems when one remembers that to work well a wheel needs a hard, level surface: without decent roads, a wheeled vehicle is as likely as not to get stuck or bogged down. Empires like those of the Romans or the Persians built great roads. But for the majority, the world stayed small.

However, the Industrial Revolution had a marked effect on transport. It was not so much that there was a need – but there were new opportunities: markets for faster, more efficient forms of transport. The questing of the engineers who were fascinated by the potential of the steam engine, together with the wish of entrepreneurs to take advantage of this new market spurred on the development of the railway in the nineteenth century. Commercial demands speeded up the process of change, and the arteries of transport grew rapidly. Not only railway networks, but also better roads brought places effectively nearer than they had been before.

The process continued with the development of aircraft. Here the story was slightly different. Business people were slower to see the commercial possibilities of air transport. The research work was done by individuals, who were convinced of the potential of what they were creating and fascinated by the idea of the human conquest of the skies. True, the Wrights themselves were businessmen. But their enterprise was more of a technological quest – at least at the beginning – than a financial one.

There are similar stories in the field of message-carrying. Writing, for example, seems to have begun as a method of keeping accounts: innovation and finance went hand in hand. Alexander Graham Bell, on the other hand, made a fortune from his most famous invention, the telephone, but had no idea of what he was instigating when he started work. It may be that history is right to see Bell as one of the greatest inventors, displaying an admirable combination of qualities: care for people, business awareness, and the sort of vision that can look out into the distance and find a new way of making the world seem smaller.

BOATS AND RAFTS

Learning to travel on water was highly significant, as it allowed humans to explore new environments and exploit different natural resources

We live surrounded by water. It is not simply that the oceans cover some 70 per cent of the Earth's surface; there are also the many lakes, rivers and streams. Our ancestors often found themselves settling near these, not only because fertile land is often to be found near rivers, but also because the waterways themselves provide a welcome source of water and fish. Early on, people realized that it would be to their advantage to go beyond the riverbank and into the water itself. Here they would find a more plentiful supply of fish, and water could also provide a means of transport.

No one knows when the earliest boats were made. Human occupation of Australia had begun by about 50,000 years ago, and by that time the island continent was already separate from south-east Asia. So people presumably travelled there using some sort of boat or raft.

Perhaps simply by watching a branch floating along a stream, people realized that they could develop buoyancy aids. And so our first ventures out into the water were probably made holding on to a log, drifting along a river, with the occasional use of a hand or an arm to steer. In some parts of the world it was found quite early on that a container full of air was particularly buoyant. And so floats made of inflated animal skins, sewn tight together to keep in the air, and also clay pots began to be used.

Rafts

The earliest proper water-going craft were rafts, craft you sail *on* (as opposed to *in* a boat). Such rafts were made by binding together such unlikely floats as animal-skin bladders, which could support a large platform capable of carrying people or cargo. An Ancient Assyrian relief of *c.*700 BC shows such a raft made of a wooden framework buoyed up by numerous animal skins inflated and lashed together. Etruscan carved gems of the sixth century BC show rafts made up of a number of pots.

Rafts in the ancient world were kept afloat in a variety of ways – with skins, pots, or logs. All of these methods had also been used to provide simple swimming floats.

If such craft seem bizarre today, they had their advantages to people of the ancient world. Pots were easy to make and particularly suitable in waters where there were few rocks. Skins could be deflated and carried if necessary. The latter was an important advantage, since early sailors could often only drift with the current. So they might set off downstream with their cargo and return, with deflated skins, by land. Early images of skin-rafts often show a donkey on board, a useful beast of burden to have for the return journey.

One area where different conditions prevailed was that of Ancient Egypt. In the River Nile, the Egyptians had an ideal waterway, one which was continuously navigable for some 750 miles (about 1,200 kilometres) and along which one could drift with the current or sail back with the prevailing wind. In Egypt, timber was scarce, but another suitable raft-building material was discovered – the reeds which grew along the banks of the River Nile.

And so the first Egyptian craft were made of bundles of reeds lashed together. By the fourth millennium BC these had become quite sophisticated. They were elegantly pointed at either end, they could be paddled along, and they were steered with oars. These were all important developments, but the most important of all was a response to the prevailing wind on the Nile: many Egyptian rafts were equipped with sails, making their sailors truly independent of the river's current.

Coracles and dugouts

For many purposes rafts were quite adequate. But the added benefits of a boat – in which crew and cargo could be kept dry – were soon realized. Boats too could be very simple; a light, wooden framework over which hides could be stretched was probably one of the earliest forms. Like an Assyrian bladder-floated raft, such a boat could be made with simple technology

Traditional coracle or quffa, with wooden frame and skin covering

(flint knives and bone needles). And it shared the raft's flexibility: both large and small examples were made, and the small ones could be made easily portable, so that you could carry your craft on your back if necessary. As with the raft, the first illustrations of such vessels come from Assyrian reliefs, the earliest of which dates to the seventh century BC. Similar boats have long been used much further north. Julius Caesar saw them in Britain – and they had probably been used there for a long time before the Romans arrived. They are still used in remote parts of the British Isles, where they are known as coracles, and in the Middle East, where they are called *quffas*.

Another simple form, perhaps the earliest of all, was the dugout canoe. This was a tree-trunk, shaped and hollowed out to form a boat. Again, such a craft could be made with the most basic technology – the wood could be hollowed out with a flint knife or by the use of fire. Archaeologists have found dugouts dating to as far back as about 6300 BC and a paddle dating to around 7500 BC preserved in a peat bog at Star Carr in England. Dugouts continued to be used in parts of the ancient world where there was a plentiful supply of timber. There were also the bark canoes, so important in North America, simple half-cylinders of bark taken straight from the tree, and the ends blocked up with clay. Such craft have survived in some parts of the world until modern times.

Plank boats

If you wanted to make a larger boat, a different method of construction was required. People probably started to expand their craft by adding wooden planks to their dugouts, building the sides upwards and outwards until the original tree-trunk became a keel. We have little information about this development, or about how the planks were joined; they may well have been attached by sewing – we know this was done in India and probably in the Mediterranean world, too.

The best evidence about plank boats in the ancient world comes from Egypt: some of the actual hulls have survived, including the magnificent solar boat buried next to the Great Pyramid of Cheops at Giza. The Egyptians had a different approach, basing the shape of their wooden boats on their reed rafts. The vessels had similar curved ends and a two-legged mast (originally introduced because it is difficult to secure a fixing for a

Shipwrights at work

Trees were not plentiful in the Nile Valley: the only local tree of much use for shipbuilding was the acacia, which provided only short planks of timber. Bundles of reeds provided a popular material for rafts. When the Egyptians started to travel, they found tall cedars and junipers among the hills of Lebanon and Syria. The latter were used for larger craft, but obtaining them involved much labour and expense, so small boats had to be put together from many small planks. Ancient Egyptian shipwrights had stone axes and adzes to cut and shape their timbers. They used dowels, mortice and tenon joints or clamps to join them together. A twisted rope allowed them to pull the deck timbers into a curve to match the planks of the hull.

single pole in a reed raft). And since a reed raft had no internal strengthening framework, neither did Egypt's first wooden boats. They were constructed by starting with the keel, and mortising or clamping subsequent timbers to it and to each other, working outwards and upwards until the required shape was achieved. At this point the shipwright added cross-beams to brace the structure and to provide support for the deck.

The Egyptians used this 'skin of planks' method of construction to provide a variety of craft, both river boats and sea-going vessels, from vast barges capable of transporting tall stone obelisks, to small craft for fishing or personal transport. The river boats could be propelled by paddles, oars or poles. Some also carried tall oblong sails, at first with two-legged masts, later with pole masts. Sea-going ships were the same shape and were built in the same way, but had some extra features, including a rope truss, a length of rope running the whole length of the deck, connecting bows and stern. A wooden lever was pushed through the fibres of the rope so that it could be turned to tighten the truss, keeping it in tension and stopping the bow and stern from sagging. Such a boat is shown in a relief that dates to about 2450 BC.

Such elaborate boats probably did not develop quite so quickly in those areas where conditions were less favourable for river transport. In Mesopotamia, for example, where there were also advanced early civilizations, rapids, shallows, and adverse winds made the Tigris and Euphrates less than ideal. Here, craft like the *quffa* prevailed, together with wooden vessels rather smaller than their Egyptian counterparts.

The coming of the galleys

At some point after this time, someone in the Mediterranean or the Middle East must have begun to build ships based on a wooden framework rather than on a 'skin of planks'. It is difficult to say exactly when this happened, but the earliest evidence seems to come from Egypt itself. There are two tomb paintings of about 1400 BC, showing groups of Syrian merchant ships. These ships must be sea-going vessels, but lack the rope truss of the earlier Egyptian vessels. The implication

is, therefore, that their hulls had a strong enough internal structure to hold them together at sea, in other words, that they were based on a wooden framework.

Another Egyptian picture, this time a relief of a battle, displays similar evidence for warships. This relief shows a battle between the Egyptians and the so-called 'Peoples of the Sea', who invaded the area around 1200 BC. Both sides have ships of sturdy construction, without rope trusses, once more suggesting that they were frame-built. The Egyptian warships are galleys, powered by both a sail and a bank of oars. Galleys were to become the standard fighting ships of the ancient world, but these Egyptian ones look very different from the famous warships of Classical Greece and Rome. For the parentage of these ships, we have to look further westward, to the Aegean.

The Aegean civilizations, such as that based at Mycenae, certainly had well-developed ships. Unfortunately, the evidence for what they were like is quite scanty. Paintings on a clay box and a vase of about 1200 to 1100 BC from Mycenae give us some idea. The vessel on the vase, in particular, with its single oblong sail, bank of oars, and curved sternpost, looks like the ancestor of the galley that was to become the favoured fighting ship in Greece and Rome.

In the Classical world, galleys were used as fighting platforms: they carried marines who, when their ship interlocked with an enemy vessel, would fight what was effectively a floating land battle. These tactics continued unchanged for hundreds of years, with some exceptions, such as the Vikings, who used their fast, light ships to carry troops to new territories where they could go ashore and ravage the countryside. The next major development would be the evolution of the galleon, the full-rigged ship that developed in all sorts of different configurations, and as early as the fourteenth century carried firearms. By this time it was clear that the ship, in all its forms, was one of humanity's most important and enduring inventions, whether it was used to go in search of food, for transporting people or goods, for exploring new territories, or in sea battles that changed the political structure of the world.

THE COMPASS

Directional devices that allowed long-distance travel were immensely important in the spread of ideas from one culture to another

The magnetic compass is one of the most important navigational aids; it is difficult to imagine navigation without it. But it appeared in the West only at the end of the twelfth century AD, and was not widely adopted until the thirteenth century. It works because the Earth has a magnetic field rather like that of a bar magnet, with magnetic poles near its geographic poles. The magnetic pole in the North attracts the 'North Pole' of the compass, so the compass needle always points north (or nearly so, since the Earth's magnetic and geographic poles do not coincide exactly). There is evidence for the early knowledge of the properties of naturally magnetic lodestone in a number of places. Ancient Egypt is one possibility; the Olmecs of Ancient Mexico may also have had compasses as early as 1000 BC. But the earliest hard-and-fast evidence for the appearance of the compass comes from China.

Origins in the East

The earliest surviving written treatment of the compass is in a text of the fourth century BC, *The Book of the Devil Valley Master*, possibly written by the philosopher Su Ch'in. The book asserts: 'When the people of Cheng go out to collect jade, they carry a south-pointer with them, so as not to lose their way.' What is interesting in this text is that the compass is already being used practically, to help people find their way. There is another mention in a slightly later work, *The Book of Master Han Fei*, confirming this use. And the phenomenon of lodestone may date back to long before these first written accounts.

Travellers may not have been the first in China to use the compass. In China there is an ancient tradition of geomancy, the art of designing buildings and towns so that they are in harmony with the Earth and its natural forces. A geomancer was usually consulted before a house was built, and a compass became an essential geomancer's tool for finding directions. The use of compasses for these 'magical' purposes is one reason why the pointers of early Chinese compasses were often shaped like spoons. This shape symbolized the Great Bear constellation, bringing to mind astrology as well as direction.

Chinese mariner's compass

We do not know for sure when the compass was first used on board ship. Several Chinese texts of the early twelfth century AD (before the compass got to Europe) mention the mariner's use of the device. By this time it had been realized that iron or steel could be magnetized by being rubbed on a piece of lodestone, so spoon-shaped pointers were abandoned, and a magnetized needle floating in a vessel of water was often used.

The compass comes to the West

It was probably in the later twelfth century that knowledge of the compass came to the West, but it is still not known exactly how this knowledge was transmitted. Unlike many Chinese inventions, it seems not to have come via the Islamic world, where the earliest accounts of compasses are slightly later than those in western Europe. The first European mention was in a work produced in 1187 by the English writer Alexander Neckham. This was followed up in 1269 by a proper study of compasses and the whole phenomenon of magnetism, *De Magnete*, by the French scientist Pierre de Maricourt. By this time the floating needle had been largely replaced by a magnetized needle balanced on a pivot inside a box covered with glass. The points of the compass were marked on a card in the bottom of the box, so that they could easily be read.

At around the same time the sand glass was developed in Europe. This allowed one to measure equal intervals of time. Thus, if the speed of a ship could be estimated, it was possible to make a good guess at the linear distance one had travelled. The compass and sand glass soon led to more accurate charts, and when all three were used in combination it was possible for sailors to work out their position without the aid of the sun or stars.

The result was that ships could operate at times of the year and in conditions when they had previously been confined to harbour. In addition, better charts were produced. This revolution in navigation both made shipping safer and boosted trade: it saved lives and made fortunes.

Charts and compasses

The earliest known compass from China consisted of a spoon made of lodestone mounted on a bronze plate. The handle of the spoon pointed south. Later, the traditional Western compass with its pivoting needle evolved. It transformed sea travel and led to other important navigational developments. One result of the use of the compass at sea was the improvement in charts. Portolani charts, with their star-shaped lines showing bearings from each major port, were first drawn in the thirteenth century. They soon covered the whole of the Mediterranean coast.

M
VM
LVS
CAN
SECV
N
DVM
CRI
TERCI
QVAR
T

THE WHEEL

Although the Romans propagated the idea that the Celts were barbarians, Celtic wheel technology was superior to their own

Nowadays, the wheel is such a common phenomenon that it is difficult to imagine transport without it. But it has not been so ubiquitous for much of human history. In prehistoric times, when uneven tracks and pathways were the main arteries of human travel, wheeled vehicles would have been next to useless. So, if someone wanted to transport goods that were too heavy to carry, their first resort would probably have been to a pack animal. Mules, llamas and camels have been used in this way for thousands of years, and still are. Even where horses were available, incidentally, mules were generally preferred. They are notoriously stubborn, but also admirably patient and sure-footed, and often easier to train for load-carrying than horses. The thicker skin of the mule is also more resistant to the chafing that can occur when carrying a heavy load, and provides better protection from the cold. With such beasts of burden, it was usually just a matter of balancing the load on the animal's back. No wonder this form of transport is so widely used around the world, even today.

Primitive vehicles

But pack animals were sometimes more trouble than they were worth. They could not negotiate narrow pathways, they had to be loaded and unloaded, led, fed, watered and guarded. The invention, sometime in antiquity, of panniers, overcame the problem of balancing the load. But even so, human porters were often preferred for all but the heaviest loads.

And there were ways of making the human-borne load easier to carry. One possibility was to put the load on rollers made from tree trunks and push it along. Such a method may have been used for moving large building stones over short distances. But it shared some of the disadvantages of the wheel. Another way was to use a simple sled, whose runners would glide easily over snow or desert sand, and which would cope reasonably well with less even surfaces. The sled also had the advantage of being relatively simple to make.

In other regions, people evolved the travois. In its typical form, this consists of two wooden poles, joined to each other at one end and diverging towards the other end to accommodate some sort of netting or framework to take the load. The narrow end is pulled along by a person or animal, while the load trails along on the ground behind. For some nomadic peoples the travois had a special advantage: its two poles could be taken apart at the end of the journey and used as supports for tents. It was especially popular in the Americas, where there were no wheels until the Europeans arrived.

Mules pulling a travois

Both sled and travois could be pulled by humans or animals, the latter varying according to what was available locally – dogs or reindeer in northern latitudes, mules or horses further south. So, both forms of transport could be well adapted to the needs of people in a variety of places. Such simple aids to transport were used for thousands of years. Sledges are still recognized as excellent vehicles in the polar regions; they were in use in Europe by 5000 BC, and they had probably been invented independently in other parts of world by then, too.

The coming of the wheel

But in Mesopotamia, something happened that was to change transport for ever. On a clay tablet of about 3500 BC from Erech in Mesopotamia, there appears a pictographic image of a sledge mounted on wheels: the wheeled vehicle had arrived.

As befits an image used symbolically in this way, the sketch is very simple and conventionalized. Yet it suggests several things about the early development of wheeled vehicles.First, the date indicates that the potter's wheel and the wheel for transport were developed at around about the same time. Second, the curved runners above the wheels show that the first carts were likely to have been adapted sledges. Third, the picture is a side view, with two wheels visible, so this must have been a four-wheeled vehicle. Fourth, the wheels appear to be solid.

By 2500 BC we have the remains of actual vehicles that confirm some of these points. The great royal tombs of the city of Ur, excavated by Sir Leonard Wooley in the 1920s, were richly stocked with grave goods, providing much information about Ancient Mesopotamian culture. Among these grave goods were several vehicles; their remains, together with those from other tombs, and ancient model carts from Syria, Babylonia, and the Indus valley, give us a fair impression of what these vehicles were like.

The wheels were normally made of wood, and most were made up of three parallel planks joined by a pair of cross-pieces running at right-angles. The central

Types of wheel

Solid wheels were normally made of wood. Sometimes, in timber-rich areas such as Scandinavia, a single plank (1) wide enough to make an entire wheel was used. But more often the three-plank or tripartite design (2) was adopted – it was strong, easy to make, and did not require large pieces of timber. Where there was no wood at all, stone wheels (3) were sometimes used. All of these wheels were heavy, and wheelwrights experimented with removing timber from the wheel to make it lighter. Sometimes, the basic tripartite design was kept, but with some wood cut out of the middle (4). Sometimes, the cross-bar design, with an outer rim and strength-giving struts, was adopted (5). Finally, the spoked wheel (6) was invented. The most common of all these designs were the tripartite and spoked wheels.

1

2

3

4

5

6

plank would be thicker at the centre, and this area would be carved to form a hub. It is not clear from either the remains of full-size carts, or the surviving models, whether they had fixed or moving axles. It is interesting, though, that wheels from different parts of the world should have a similar structure – right down to the usual number of planks being three. This suggests that the wheel was developed in one place and then gradually spread around the Middle East – and eventually beyond. This theory seems to tie in with the dates of the evidence for use of wheels that has been discovered. The dates get later the further one travels from Mesopotamia. Wheels came to Mesopotamia around 3500 BC, were in use in Assyria by 3000 BC, were known in the Indus Valley and in Central Asia by 2500 BC, and arrived in Crete and southern Russia by 2000 BC.

Although these early wheels were very simple, they could have arisen from even more basic antecedents – it was once thought that wheels derived from rollers. Someone, so the theory went, took a 'slice' of a tree-trunk roller and attached it to a vehicle to make a wheel. This might well have been tried, but the resulting wheel would not have been very strong, since the timber would probably have quickly split along the grain. So if wheels did derive from rollers, the first ones would have been short-lived. Few places, especially in the Middle East, had trees large enough to yield single planks big enough to make wheels. And so they had to be constructed from several planks joined together.

Wheels also depended on metal technology. For a start, the planks to make a wheel had to be cut accurately, for which task a saw was really necessary. In addition, the edges of the wheel were subject to wear. The best solution was to reinforce them with metal, the original way being to drive in copper nails around the circumference. (According to some authorities, it is possible that such nails were also used to attach leather tyres.) This dependence on metal suggests one reason why the wheel was not adopted earlier.

The first wheeled vehicles

On what kind of vehicles were these early wheels actually used? This is a difficult question because of the nature of the evidence. Many of the remains of early wheeled vehicles come from royal tombs, where they were used on hearses to carry the deceased to their final resting place. It is therefore tempting to suppose that this was the main first use of the wheel, and that wheeled vehicles were at first luxury items, the preserve of kings and queens. So it may have been, but this should not blind us to the fact that there may have been many wheeled vehicles used in humbler contexts, which have not survived.

Another type of vehicle in which wheels were used quite early, and in quite large numbers, was the war chariot. Such vehicles are represented on items such as a famous inlaid box from the tombs at Ur. Rather like the royal hearse, the chariot was a prestige vehicle. It was an early example of the tendency for resources to be found for new developments in military technology, which may then be passed on to other areas.

So, although the design of early solid wheels was remarkably similar across a wide geographical area, the vehicles on top of them varied considerably. This was perhaps partly to do with the fact that wheels were often applied to vehicles that already existed, as we have seen above, with the Mesopotamian sledge on wheels. Thus, early wheeled carts and chariots often had a rather makeshift character. There is even evidence that the axle was sometimes attached to the vehicle body by means of simple leather straps.

This 'cobbled-together' approach had an important advantage. At the time of the first wheels there were no decent roads, and the ground the vehicles had to negotiate was often difficult and uneven; sometimes there would be rocks or rivers in the way. It was useful to be able to take a cart to pieces and carry it over such obstacles,

Making wheels

We have important evidence about early methods of wheelwrighting from the tomb paintings of Ancient Egypt. These usually show felloes made in several sections, and fixed to the four spokes with mortise and tenon joints. Sometimes the felloe was made in one piece, by heating and bending a long strip of wood.

The Egyptians also began to use metal in wheels, either by sheathing wooden wheels in bronze, or by making the entire wheel from metal. Neither of these solutions made the wheel stronger or more durable. These decorative wheels were used in vehicles that were designed to be buried in royal tombs, for the use of the deceased in the afterlife.

in the same way that early sailors often had to carry their rafts over rapids. This also accounts for the fact that the vehicle body was often quite lightweight – either of thin planks or of wickerwork. Coverings could be fabric or skins.

At the front of the vehicle there was usually a single pole to which could be harnessed two or four animals – oxen for heavy transport, asses to pull chariots. Horses appeared around 2000 BC, and the remains of bits (i.e. mouthpieces) appear in the archaeological record soon after this date.

The development of the wheel

The main drawback with solid wheels is their weight. This is a particular problem with chariots, which need to be fast and manoeuvrable. The spoked wheel probably evolved because of the need for lighter wheels. In some places, wheelwrights tried to lighten their wheels by cutting out sections from the part of the wheel between the rim and the hub. But this had to be done with care if the wheel's strength was to be preserved, and sometimes led to the inclusion of reinforcing pieces. A better solution, if the wheelwright had the skill, was to make a spoked wheel. This consisted of a central hub, several spokes, and the circular outer part, usually made up of several sections called felloes. To begin with, four-spoked wheels were quite common. These were seen in places as far apart as Egypt (around 1400 BC) and China (about a hundred years later).

The wheel would eventually transform transport, but it would really do this only when road networks were extensive enough, and good enough to make wheeled vehicles truly effective. This, above all, is probably the reason why there is such a long gap between the appearance of the wheel, in around 3500 BC, and its establishment in regular and common use up to 2000 years later. But wheeled vehicles and good roads together would ultimately have a great impact on trade, and on the movement of people and artefacts around the world.

The wheel had a much more immediate impact on warfare. As we have seen, the chariot was adopted much more quickly than the cart. As used by the Ancient

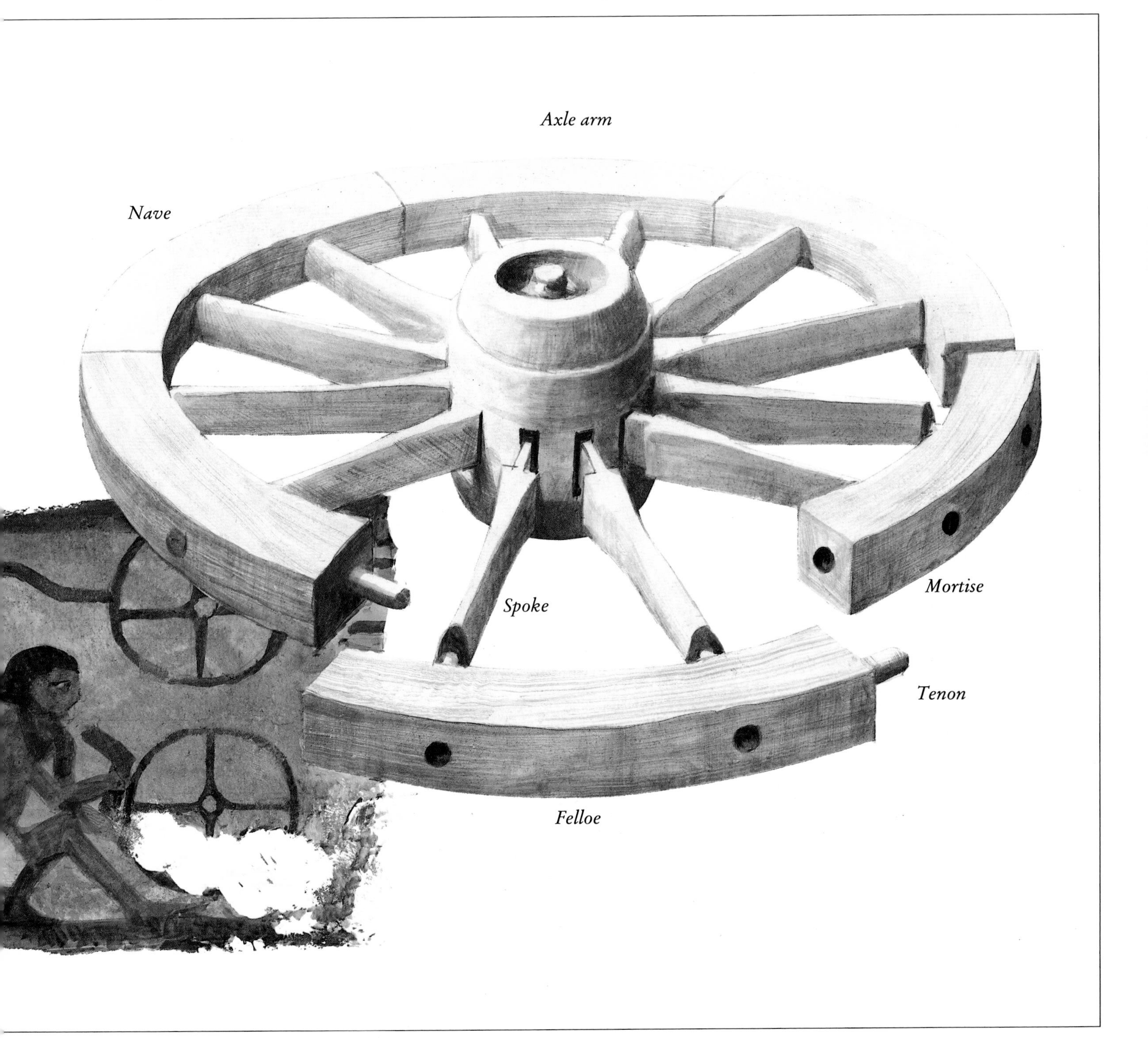

Egyptians, the chariot allowed archers to move around at great speed, the vehicle acting like a highly mobile firing platform. As used by the Hittites of Anatolia, the chariot itself became a weapon, capable of driving into a mass of foot soldiers and breaking them up in confusion. Such an effective fighting machine left its scars on the ancient world; it played its part in the success of peoples like the Hittites and the Assyrians in their attempts to dominate the world. But the chariot also had a positive effect. The developments in spoked wheels that were at first used only in chariots eventually found their way into civil vehicles, to the benefit of farmers and merchants alike.

The Greeks and after

By the time Classical Greek civilization reached its peak there were two types of spoked wheel in regular use, one usually seen on carts, and one on chariots. Either fixed or rotating axles were used on both types of vehicle.

The cart wheels of the Ancient Greeks were generally of the cross-bar type. They had a rim made up of two felloes, and a cross-bar that extended across the diameter from one outer edge to the other. This was strengthened with plates in the middle of the wheel to form the hub.

Greek chariot wheels usually had four, six or eight radial spokes, which were often turned on a lathe. The felloes were probably made up of several sections, and it is possible that the whole wheel was sometimes further strengthened by the addition of a metal tyre. Further strengthening was probably also necessary at the points where the spokes joined the felloes.

Celtic carts

During this period solid-wheeled vehicles were also still being used. Indeed, there are illustrations of solid-wheeled carts from northern Europe dating from as late as AD 300. However, the Celtic peoples of northern Europe had much more sophisticated wheels, and some have been preserved. Some of the most impressive belong to a group of wagons buried in a peat bog at Dejbjerg, in Denmark. The wheels of these vehicles had twelve spokes of hornbeam, single-piece ash felloes, and hubs of oak. The wheels were adorned with bronze fittings. The most interesting aspect of their design is what seem to be wooden bearings. The hubs are grooved, apparently to take wooden rods that turned between them and the axle. There is some doubt as to how effective these bearings would have been, and the wagons, probably made specifically as grave goods to be buried with their owner, most likely never had to carry a heavy load over

Detail of Scandinavian cart

a long distance. But the very existence of the grooves suggests that the wheel-wright's craft had reached a high degree of sophistication, a fact also borne out by the complex mortise and tenon joints linking the spokes to the hubs.

Celtic wheels also often had iron tyres. These were usually placed red-hot on to the felloes and then quenched with water. This shrunk the tyres on to the wheels and simultaneously hardened the metal. The evidence is, then, that the Celts, whom the Romans thought of as barbarians, had a more highly developed wheel technology than the Romans themselves.

Medieval and later wheels kept similar designs to these early examples. No long-lasting solution to the problem of bearings was found until the eighteenth century, when the idea of ball bearings (probably devised originally by Renaissance goldsmith Benvenuto Cellini in the sixteenth century) were applied to carriage wheels for the first time.

Ball bearings were widely used from the mid-nineteenth century on. But they really came into their own with the fast speeds and high friction generated between wheels and axles in the powered vehicles of the twentieth century.

It was a similar story with tyres. Rubber tyres, originally solid, appeared in the nineteenth century. The air-filled tyre appeared as early as 1845. This was the year in which British engineer Robert Thomson filed a patent for a leather air-filled tyre.

Thomson's tyre did not catch on, but rubber pioneer John Dunlop saw the potential of an inflatable rubber tyre with a rubber inner tube. The appearance of powered vehicles like the motor car made good tyres increasingly important – both for comfort and for better grip on the road. In 1890, Charles K. Welch, another British inventor, patented the dished wheel rim and wires to help keep the tyre in place. This made the inflatable tyre more of a practical proposition. Even so, early cars and lorries made do with solid rubber tyres.

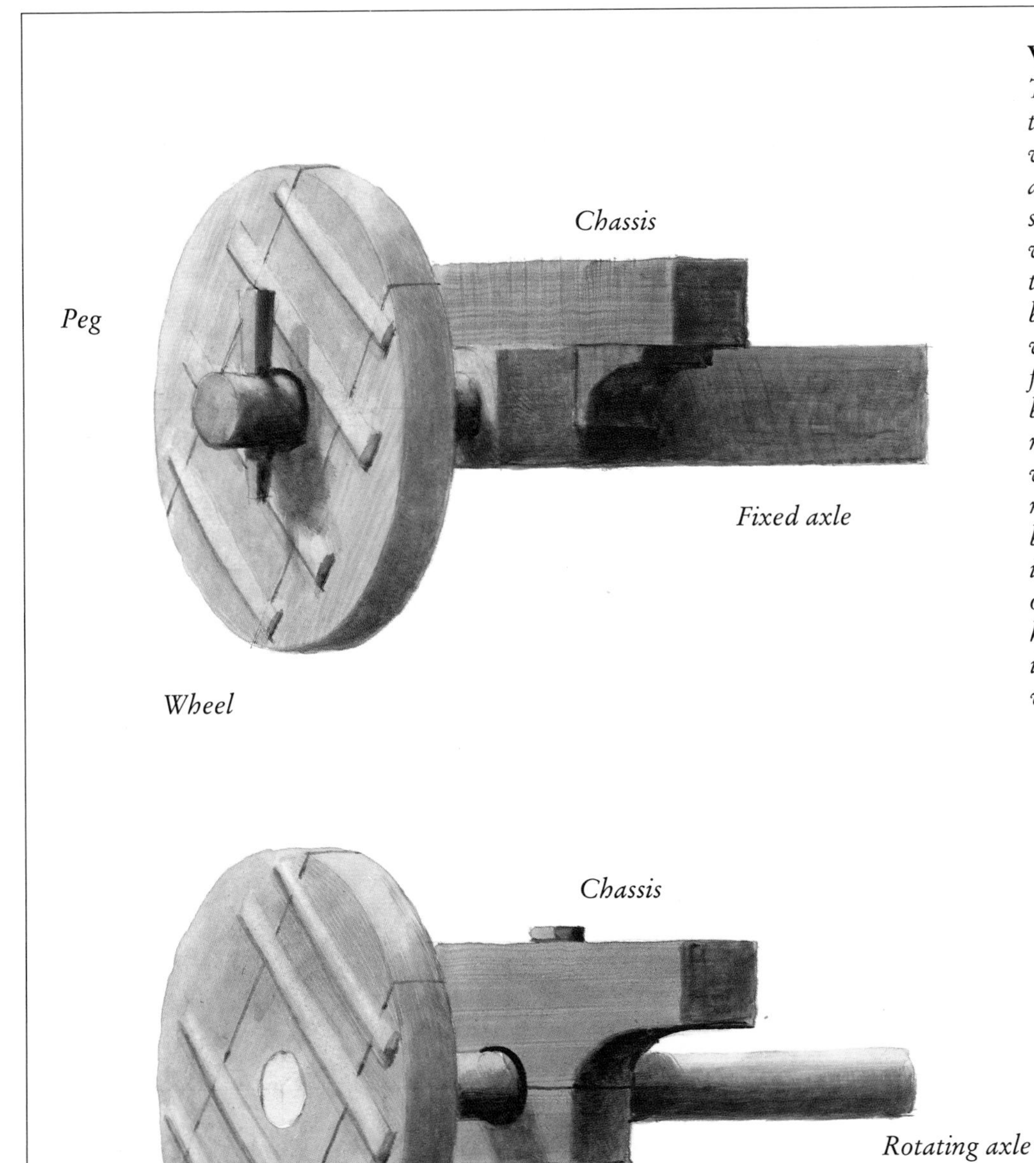

Wheel and axle

There were two basic ways of connecting the wheel to the axle. In one type, the axle would be fixed to the chassis of the cart, and the wheel would turn around a short stub at the end of the axle. A wooden pin would be passed through the stub to stop the wheel falling off. A metal washer between the pin and the hub reduced wear. Alternatively, the wheel could be fixed to the axle, and the axle turned in a bearing underneath the cart. The first method was more suitable to light vehicles, the second to heavier ones. With most of the parts of the cart and wheel being made of wood, much wear was inevitable in both types. In the fixed axle or stub design, this would occur on the hub and stub. On the rotating axle design, it would be the bearings beneath the vehicle that would show the most wear.

ROADS AND RAILWAYS

Networks of well-made roads were made by the Persians, Chinese and Romans, extending travel beyond the well-worn paths of trade routes

When people began to grow crops, domesticate animals, and settle down, travelling took on a new significance. Before, when people led a nomadic life, they travelled from one source of food or shelter to another, ranging across the landscape according to what was available and where. However, in a more settled existence, and with the possibility of using pack animals to carry goods, their movements became much more between fixed points: their own settlements and those of others. This was even more likely to be the case when people started to live in towns or cities, and to trade. So the earliest tracks and roads probably marked these movements.

Roads in the ancient world

It is not surprising, then, that the first great road-builders were the Mesopotamians. They had roads connecting Babylonia and Egypt, as well as processional routes between cities such as Assur and Babylon. These roads, or at least the later ones (*c*.700 BC) that have survived, are well made. There is a foundation of rubble and gravel, covered with courses of bricks, and topped with stone slabs. Such a multi-layered approach to road-building, to give both a strong foundation and a flat top surface, is still followed in principle today, even though the composition of the layers has changed.

Another well-made ancient road was the Persian Royal Road, in use as early as 3500 BC, and until *c*.300 BC. This was, in fact, a network of roads, beginning at Susa on the Persian Gulf and spreading northwards to centres such as Jerusalem, Nineveh, Boghazköy (the Hittite capital) and the Black Sea. The Chinese also had roads built by command of the emperors; these often included stone-paved steps up mountainsides. However, although impressive, only the more powerful rulers could keep such roads well maintained for any length of time. Other ancient peoples who built roads included those of Malta and Ancient Greece (both of whose roads had grooves or ruts carved into the top surface so that they could take cartwheels), the Indians, the Egyptians, and the Incas of South America.

Probably the greatest road-builders of the ancient world were the Romans. Their roads were admirable for their straightness as well as for their sound construction, with foundations of broken stones in mortar, and pavements of flat slabs. The Romans also had the idea of a network, connecting the diverse territories of the huge empire, and built some 53,000 miles (85,000 kilometres) of roads.

Charioteers on the Persian Royal Road

The most impressive Roman road-building achievement was the Appian Way, started in 312 BC, one of the great military roads that converged on Rome itself. The structure of the roadway was quite complex, and the foundations could be up to five feet (one-and-a-half metres) deep. On a layer of sand or mortar in the bottom of the trench were first of all laid large flat stones; next came a course of smaller stones mixed with lime; then a layer of gravel and coarse sand mixed with hot lime; finally the top surface of the road was applied: a layer of flint-like lava. The whole was cambered so that water would flow away.

After the fall of Rome, road-building went into decline; the people of medieval Europe relied on trains of pack animals rather than wheeled vehicles, so well-paved roads were unnecessary. Towards the end of the Middle Ages, better carts began to be produced again, so the demand for better road surfaces was more pressing.

New ideas in road building

The few good roads built in the Middle Ages used Roman methods, but these were labour-intensive. Consequently, engineers began to cast around for a surface that was thinner and easier to lay, one in which the ground could support the load, rather than the roadway taking the whole weight, as in the Roman types.

In France, in 1764, Pierre-Marie-Jerome Tresaguet came up with an alternative to the Roman method. Tresaguet's roadway was much simpler in design than the Roman model. It consisted of a layer of coarse uniform stones, covered by another layer of smaller, walnut-sized broken stones. The whole roadway was no more than ten inches (twenty-five centimetres) thick.

The eighteenth century also saw a number of important British road-building pioneers. The first of these was John Metcalf, who was building roads during the middle of the century. It was Metcalf who pointed out the importance of good drainage, and gave his roads ditches on either side, in addition to a six-inch

The Romans built a network of roads around the Mediterranean and across Europe. These were usually straight and well built, and more like underground walls than roads, with layers of large rocks topped with a pavement of flat stones to give a smooth surface.

(15.25-centimetre) crown in the middle to take away the water.

Another important British road-builder of this time was Thomas Telford. In fact, Telford was well established as a civil engineer, with numerous bridges, harbours and drainage schemes to his credit, before he turned to road-building at the beginning of the nineteenth century. Telford insisted that the principles of engineering were rigorously adhered to when building his roads. He had three key tenets: first, the roadway should be as flat as possible, with a maximum gradient of no more than 1 in 30; second, the stone roadway should be able to carry the heaviest loads required; third, the roadway should be adequately drained.

Telford had his own solution to the problem of a strong roadway. His roads consisted of two layers. The first was a foundation of uniformly sized large stones, placed carefully by hand. The second was a layer of irregular smaller stones with a thin layer of gravel on top. There was a slight crown of about four inches (ten centimetres). The continuous passing of metal-rimmed wheels on this sort of roadway was designed to grind up the small stones of the top layer into a fine dust. This dust worked its way into the gaps between the remaining stones, hardening and binding them together as it did so. The result, after a while, was a smooth, durable covering.

Telford's roads were very well made and lasted for years. But they were expensive to build. The hand-grading and placing of the stones took an immense amount of time. The man who saw a solution to this was John McAdam. McAdam was employed as a highway inspector – it was his job to examine the roads of a particular area and to arrange for necessary repairs. He also built some new roads. McAdam saw that a well-drained subsoil could support the weights that roads were usually required to carry. He therefore made roads with a mere ten inches (twenty-five centimetres) of surfacing, consisting of small stones laid in loose layers. The upper part of this surface was pulverized by passing traffic, and lime could be added to the resulting dust as a binding agent. McAdam had an efficient procedure for the checking and replacing of weak patches of road surface and, as long as this was done, the system proved very effective for traffic consisting mainly of heavy, slow-moving carts with wooden or metal wheels.

Roads built using McAdam's methods proved very popular until the coming of rubber-tyred vehicles. These tended to loosen the stones and create great clouds of dust. The solution was to use bituminous tar as the binding agent.

From the point of view of the number of roads built, the methods of Tresaguet and McAdam proved most influential in both Europe and North America in the nineteenth century, and the great increase in road-building seen in the eighteenth century continued in the nineteenth under their influence. Even when road transport was eclipsed by the expanding railway networks and new roads became less common, road repairers carried on using their methods.

Railways – the beginnings

The idea of the railway is quite ancient. In some early road-building civilizations, as we have seen, grooves were cut in the paved surface of the road. These were designed to take the wheels of standard-gauge carts, which were, of course, pulled

Road surfaces
What is important in a road is usually below the surface. The problems of supporting the weight of passing vehicles, providing a durable roadway, and ensuring good drainage, have been solved in many different ways.

McAdam
A cambered earth foundation lay beneath McAdam's roads. Next came two four-inch (ten-centimetre) courses of stones, followed by a top course of smaller stones, which passing wheels crushed to a smooth surface. There was a drainage channel on either side.

Tresaguet
The French engineer used heavy stones pushed into a base of earth. Above this were two thin layers of smaller stones. Retaining stones at the sides kept everything in place. Water drained away well from this type of road.

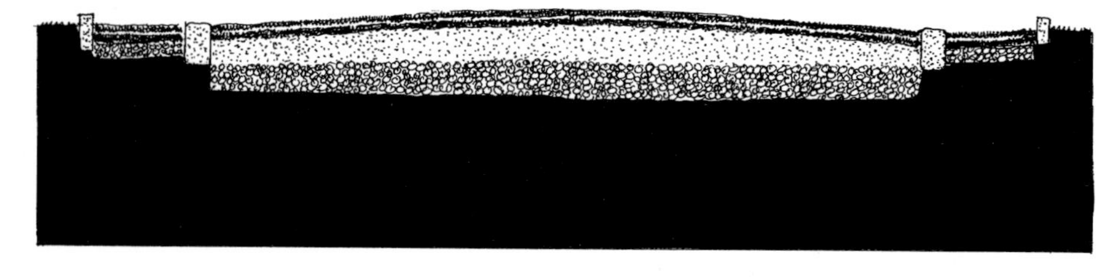

Modern roads
Many modern roads have a sub-base of granular material, followed by a concrete base and a layer of tar or rolled asphalt. This is topped by the asphalt wearing surface. There may be a hard shoulder with a shallower concrete foundation.

by horses or oxen. It seems a curious limitation on road transport now, but it must have made some sense to the ancients: maintenance could be concentrated on the grooves, the carts needed minimal steering, and a standard gauge of cart was used, which had advantages in assessing loads and taxes.

Much later, someone had the idea of running vehicles on rails. One particular use of this was in the difficult terrain in and around mines. Here, large amounts of heavy stone and ore needed to be removed across ground that was often uneven. Laying wooden rails gave a smooth running surface, which could be made level or gently sloping with relative ease. The track could also be moved easily or extended when the workings shifted, as they were doing all the time.

These simple wooden rails are the ancestors of today's railway tracks. We do not know exactly when they were first made, but they are first illustrated in a remarkable book about mining and metalworking called *De re metallica*, published by Georgius Agricola in 1556, so they go back until at least the sixteenth century in Europe. Such rails were a good solution to a pressing problem. However, they were not ideal in the long term. Wooden rails wore out quickly, even if they were made of hardwood or covered with hardwood strips. So, by the beginning of the nineteenth century, with the development of metal technology, rails began to be made of cast iron.

Steam carriages

Most mine owners were perfectly happy with their horse-drawn railways, particularly if they had the new, long-lasting iron rails. However, with the great development of the steam engine as a stationary power source at the end of the eighteenth century, some engineers began to look at the possibility of using steam power to drive vehicles.

The first to do this was the French engineer Nicolas Joseph Cugnot. As so often, there was a military motive behind the idea: Cugnot wanted to produce a carriage to pull heavy guns. In 1770 his first attempt appeared, followed by a Mark II model the following year. The latter (which still exists) is a three-wheeled vehicle with two 50-litre, single-acting high-pressure cylinders positioned over the single front wheel. These drive ratchets on each side of the axle, and are fed from a large copper boiler fastened in front of the front wheel. The gun carriage was designed to run on the open road rather than on rails, so it was in theory adaptable to a wide range of uses. But it managed speeds of between only two and six miles (three and ten kilometres) per hour, and therefore did not catch on.

Early steam carriage

Experiments of a similar sort went on in the workshop of James Watt, who had found success developing the stationary steam engine. In 1784, Watt's assistant William Murdock designed and built a three-wheeled steam car that ran on the roads of Cornwall, where Murdock was installing Watt and Boulton engines. It ran well, but Watt was not impressed. He assigned Murdock to more work on stationary engines, effectively stifling his assistant's project. It may be that Watt did not want him to produce something that would eclipse his own work.

Railway precursors

Horse-drawn railways were used to get coal and ore out of mines from the sixteenth century onwards. The earliest had wooden rails; these were followed with wooden rails reinforced with iron plates. Soon it was realized that cast-iron, and later wrought-iron, rails gave still greater strength and durability. Such metal rails were essential before heavy steam engines could be adapted for pulling trains.

One development that was necessary before efficient steam transport became a reality was the use of steam at high pressure. This had not proved necessary for stationary steam engines. Many engineers, Watt included, were wary of high pressure steam – they did not think that boilers could be made to withstand high pressures, and their caution was well founded. But there were some experiments with high-pressure stationary engines, notably by Jacob Leuupold, of Leipzig, in 1725.

Richard Trevithick

The first successful experiments with high-pressure steam were carried out much later, by the Cornish engineer Richard Trevithick. He began supplying Cornish mines with pumping engines of 100lb per square inch (6.89 bars) or more, in 1801. Lighter and smaller than equivalent Watt engines, and more economical, Trevithick's engines were well received, in spite of an accident when an operator tied down the safety valve of one of them, resulting in a fatal explosion. Engineer Oliver Evans met with similar success when supplying comparable engines in the United States.

It was to be the Cornishman who would adapt these engines for use in transport. Like Cugnot and Murdock, Trevithick began with a road vehicle. After small-scale models, he progressed to a full-size road carriage, which was first tried out at the end of 1801. The vehicle soon exploded when its driver allowed it to overheat, but not before it was made clear that it worked well in principle.

There followed forays into steam tramways, and steam engines for pulling boats on canals and powering river dredgers. It turned out to be Trevithick's tramway that pointed the way forward. Connecting Penydarren to Abercynon in South Wales, it stretched some ten miles (sixteen kilometres), and went into service on 21 February 1804. It was impressive. It could haul ten tons of iron, as well as some seventy people. But the idea did not catch on immediately.

Trevithick saw the need for publicity. Perhaps more influential people would take up the cause of the steam railway if he arranged a demonstration in central Lon-

1829.

GRAND COMPETITION

OF

LOCOMOTIVES

ON THE

LIVERPOOL & MANCHESTER RAILWAY.

STIPULATIONS & CONDITIONS

ON WHICH THE DIRECTORS OF THE LIVERPOOL AND MANCHESTER RAILWAY OFFER A PREMIUM OF £500 FOR THE MOST IMPROVED LOCOMOTIVE ENGINE.

I.

The said Engine must "effectually consume its own smoke," according to the provisions of the Railway Act, 7th Geo. IV.

II.

The Engine, if it weighs Six Tons, must be capable of drawing after it, day by day, on a well-constructed Railway, on a level plane, a Train of Carriages of the gross weight of Twenty Tons, including the Tender and Water Tank, at the rate of Ten Miles per Hour, with a pressure of steam in the boiler not exceeding Fifty Pounds on the square inch.

III.

There must be Two Safety Valves, one of which must be completely out of the reach or control of the Engine-man, and neither of which must be fastened down while the Engine is working.

IV.

The Engine and Boiler must be supported on Springs, and rest on Six Wheels; and the height from the ground to the top of the Chimney must not exceed Fifteen Feet.

V.

The weight of the Machine, WITH ITS COMPLEMENT OF WATER in the Boiler, must, at most, not exceed Six Tons, and a Machine of less weight will be preferred if it draw AFTER it a PROPORTIONATE weight; and if the weight of the Engine, &c., do not exceed FIVE TONS, then the gross weight to be drawn need not exceed Fifteen Tons; and in that proportion for Machines of still smaller weight – provided that the Engine, &c., shall still be on six wheels, unless the weight (as above) be reduced to Four Tons and a Half, or under, in which case the Boiler, &c., may be placed on four wheels. And the Company shall be at liberty to put the Boiler, Fire Tube, Cylinders, &c., to the test of a pressure of water not exceeding 150 Pounds per square inch, without being answerable for any damage the Machine may receive in consequence.

George Stephenson

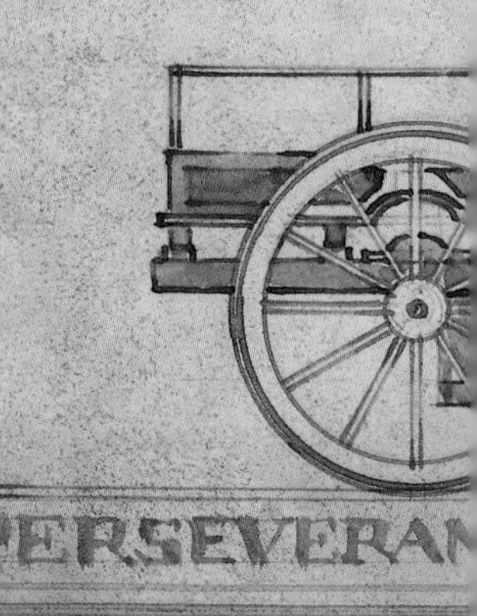

The Rainhill trials

At this crucial event in the early history of the railways, Rocket, *the locomotive that was to set the style for many more to come, was chosen and publicized. Features such as the multiple-tube boiler and tall chimney would become standard on steam locomotives in the ensuing years.*

ROCKET
NOVELTY
CE
CYCLOPED

FLIGHT

When land and sea travel became commonplace, people yearned to conquer the skies and succeeded beyond their wildest dreams

For thousands of years people dreamt of being able to fly. The mythologies of the world are populated with winged gods and goddesses, and spirits of the sky. There were also mythological bird-men, like Icarus and Daedalus in Greek mythology. Daedalus made himself and his son Icarus wings from feathers and wax, but Icarus flew too high and his wings melted in the sun's heat, sending him plunging to his death.

Not surprisingly, in this story, as in many later ones, humans looked to the birds, with their flapping, feather-covered wings, for inspiration when it came to flight. Several science-fiction writers, from the sixteenth to the nineteenth centuries, wrote stories in which men either flew on bird-like wings or were pulled into the sky by birds in some sort of harness. But for the first true aeronauts, the reality of flight was quite different, because the first humans to get off the ground did so beneath balloons filled with hot air or hydrogen, gases that are lighter than the air of the atmosphere. The achievement of making the first balloon flight belonged to the French Montgolfier brothers, and their work inspired a series of other developments in lighter-than-air flight, culminating in such wide-ranging forms of transport as the great airships, which were powered, steerable, and carried many passengers, and the balloons of recent intrepid explorers. But very different developments led to heavier-than-air powered flying machines capable of sustained, controlled flight.

Heavier-than-air flight – gliders and others

As we now know, the future lay not in flapping-winged craft, but in aircraft with wings that were fixed in position. This was not obvious to many early pioneers of flight, who looked to birds for their models. As a result, many of the most significant early developments were achieved by the pioneers of the glider, whose machines had wings that did not need to flap.

One of the most important of the pioneers was the British researcher George Cayley, who worked on flight from 1792 until his death in 1857. Cayley knew that there were several key challenges that faced any would-be aircraft designer. First, a surface capable of supporting the weight of the pilot and any other load had to be built; second, a sufficiently strong power source had to be found that was also light enough to be lifted. As Cayley himself put it, one needed to make 'a surface support a given weight by the application of sufficient power to overcome the resistance of air'.

Cayley went far towards the development of the supporting surface, but he did not live to see the light but powerful engine that was needed before aircraft were to be truly successful. He produced some extraordinary designs for a sort of vertical-take off helicopter-aeroplane. But Cayley had more success with his gliders, for which he developed the sort of lightweight, rigid frameworks and lift-giving wing-sections that later pioneers found essential. He was also aware of the importance of pilotless tests, making a full-size unmanned glider before persuading his coachman actually to take to the air in the piloted version.

Otto Lilienthal was another early aviator who concentrated on developing the glider. During the 1890s he made some 2,000 flights in a number of gliders (both monoplanes and biplanes) that he made himself. His gliders eventually enabled him to make flights of up to 1,000 feet (300 metres) in length. He was particularly interested in making his aircraft as controllable as possible. First he tried controlling the machines by moving his body-weight around. But this did not give him the range of effects he needed, so he started to experiment with moving

Mythical flights
There are many ancient stories about flying machines, like this bird-powered craft flown by an Indian prince.

Airship
In the seventeenth century Francesco Lana designed a flying ship. Lift was meant to come from four evacuated copper spheres.

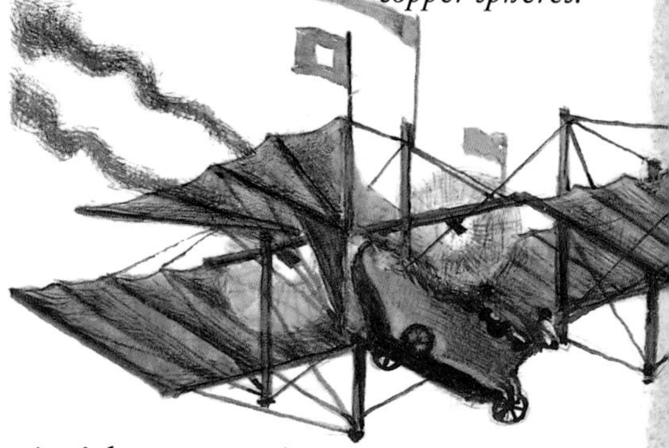

Aerial steam carriage
William Henson built models of a steam-powered flying machine that might have worked. The wing had the correct aerofoil section.

The dream of flight
Before serious experimentation got going in the late nineteenth century, many people had speculated about human flight. Most of their ideas look bizarre today, but at the time, with only observations of birds on which to draw, they looked much more practical.

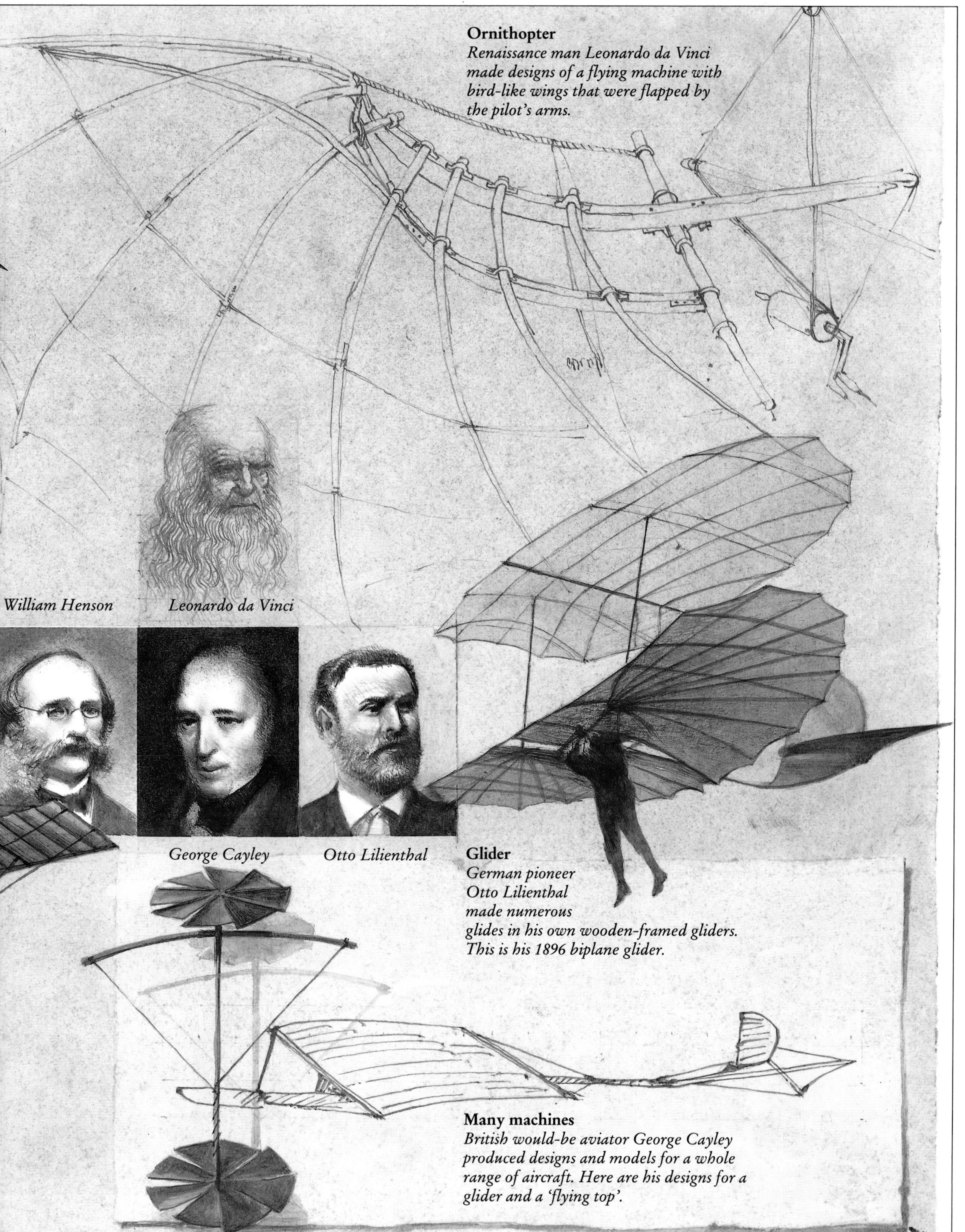

Ornithopter
Renaissance man Leonardo da Vinci made designs of a flying machine with bird-like wings that were flapped by the pilot's arms.

Glider
German pioneer Otto Lilienthal made numerous glides in his own wooden-framed gliders. This is his 1896 biplane glider.

Many machines
British would-be aviator George Cayley produced designs and models for a whole range of aircraft. Here are his designs for a glider and a 'flying top'.

elevators. In 1896, while making a flight on one of his gliders with such an elevator, Lilienthal spun out of control and crashed. His injuries proved fatal.

Lilienthal's work was taken up by French engineer Octave Chanaute. He continued the search for a better control system, as well as developing the truss-based structure that would be typical of early aeroplanes. He also recorded many of his flights in series of still photographs, gaining invaluable evidence of the effects of his structures and control surfaces.

Because of the scientific way in which Chanaute recorded what he did, and because he worked in America, his work proved very useful to the Wright brothers when they came to build their first aeroplanes. Indeed, the Wrights read everything that Chanaute wrote, and they became regular correspondents with the Frenchman.

Powered-flight pioneers

Even before Chanaute's work in the USA and Lilienthal's efforts in Europe, another European got very near to powered, controlled flight. He was the French engineer Clement Ader. In the 1880s, he built his first aeroplane, a large wooden-framed, fabric-covered, steam-powered contraption based on the wings of a flying fox and called the Eole.

The craft was a qualified success. According to Ader's own account, it left the ground during a test in 1890, flying for a distance of some 160 feet (50 metres), a distance that apparently might have been exceeded had it not been for the small size of the manoeuvering area set aside for the experiment. Unfortunately, the aircraft was not thought stable enough to continue with the tests. Ader was later given funds by the French government to continue his experiments. He produced a modified aircraft, known as Avion No III, but the success of this was also limited.

Another inventor who came very near to the goal of human flight was Samuel Pierpont Langley, Secretary of the Smithsonian Institution in Washington. He began modestly, with models. At first these were quite simple devices, powered by twisted rubber bands: they allowed Langley to test different wing-shapes and configurations.

Next, in 1896, Langley created a model aeroplane powered by a tiny one-horse-power steam engine. This was a resounding success. Langley launched it from a catapult device, on a platform over water to minimize the damage when the model landed. The aircraft rose, and, under the power of its steam-driven propeller, made several large turns above the water before running out of steam. When the engine cut out, it glided gracefully down to the surface of the water and was immediately ready for another flight.

Langley wanted to build a larger craft, capable of carrying a pilot. This aim was shared by the US War Department, whose members heard of Langley's work and were interested in using flying machines for war. The fact that the USA was at war with Spain over Cuba at the time gave an added impetus, and at the end of 1898 Samuel Langley was given a government grant to develop a machine that would be capable of carrying people.

It seemed that the time was ripe for Langley. But there was still a great deal of work to do. In collaboration with a young engineer called Charles Manly, Langley produced an internal-combustion engine to power the aircraft. Next, he built a one-quarter-scale aeroplane, which flew successfully, suggesting that the full-scale version would work too. By October 1903 the full-size flying machine was ready to test. The craft was launched from a catapult device mounted on a houseboat in the Potomac River. The first test, in October 1903, ended in failure, apparently because a stay caught on the landing equipment on the boat. In December the experiment was repeated. Again the machine failed to fly, when the rear wing structure collapsed and the aeroplane fell into the river and broke up. Langley was devastated, and subjected to harsh criticism from the press. His funds had run out and he withdrew from the field of aviation. But it was probably only the defective launching equipment that sealed the doom of his aeroplane.

The Wright brothers

Meanwhile, in Dayton, Ohio, two brothers who ran a bicycle business were pushing forward with what would turn out to be decisive research. Wilbur and Orville Wright were fascinated by flight. They read everything they could find about the work that had been done so far, including the works of Chanaute, Lilienthal and Langley, and by 1899 they were making their first experimental model.

They began with a kite, which they flew to test the stability of basic wing structures, and to assess the lift and drag these produced. The kite had two wings with an elevator in front. The ends of the wings could be bent to keep the kite balanced laterally. Already some of the basic features of the Wrights' final biplane design were emerging. By the following year they had made a full-size glider based on their kite design. First they flew it as an unmanned kite, then tethered, before

Ballooning
The first human flight was in a lighter-than-air device – a balloon. Its invention was the result of the inspiration and effort of two brothers, the Montgolfiers, of Annonay, France, and of another Frenchman J. A. C. Charles. It was Joseph Montgolfier who first thought that it might be possible to store smoke, with its rising property, so that it could be used to make objects rise. He made a small silk bag, lit a fire beneath it, and watched it rise to the ceiling.

Soon Joseph and his brother Etienne had made a larger balloon some thirty-five feet (10.7 metres) in diameter, made of fabric strengthened with paper, and held together with buttons. In June 1783 they were ready to try it out. They held it over a fire, it filled with smoke and hot air, rose, and travelled over one-and-a-half miles (about 2.4 kilometres) before landing. After a flight with animals on 21 November 1783, another Montgolfier balloon, this time with two friends of the inventors on board, took off from the Bois de Boulogne and flew for nearly ten miles (sixteen kilometres), reaching an altitude of 320 feet (900 metres).

By this time Charles had begun to experiment with balloons filled with hydrogen, a newly isolated lighter-than-air gas. His first balloon was set off in August 1783. By 1 December in the same year a human-carrying flight had taken place. The ease of use of the hydrogen balloon was admired (the pilots did not have to work continuously to tend a smoky fire), and hydrogen balloons became the norm.

It was not until 1852 that Henri Giffard's large, steam-driven airship made powered, steerable flight in a lighter-than-air machine a reality. During the second half of the nineteenth century, it seemed that the airship would provide the lasting solution to the question of flight. But vulnerability in bad weather led to disasters, culminating in the explosion of the Hindenburg *in New Jersey in 1937.*

The Wright brothers' achievement
Wilbur and Orville Wright made their great achievement is a series of steady steps. They began with kites (top), *designed like gliders but flown unmanned. When they were sure of the design, they progressed to manned glides* (middle). *Only after many tests like this did they attempt powered flight* (bottom). *Soon they were confidently making free flights as they learnt how to control their aircraft.*

making free, piloted flights. With this glider they found that they could travel over 150 feet (50metres).

However, the brothers were not satisfied with their glider, and determined not to fall into the trap of progressing to powered flight before they had perfected it. They saw that the three problems for human-carrying flying machines were, first of all, to make adequate sustaining wings; second, to create the right power plant; and third, to balance and steer the machine in flight.

The Wrights mounted a concerted attack on all three of these problems. It was particularly the third challenge, of steering and control, that had been the downfall of previous researchers, and which needed the most work. The Wrights were frustrated by the lack of reliable data on how wings and related structures behaved, so they set about working out the figures for themselves. In order to do this they even built their own wind tunnel, testing numerous experimental wing sections and propellers.

In August and September 1901 they built their Glider Number 3, based on all their findings in the wind tunnel. This had the elevator of the previous gliders, but also had a double fixed tail fin, later to become the rudder. They found that Glider Number 3 did not fly too well until they adapted the fixed tail fin to make it steerable, allowing banking and yawing. Suddenly, the question of stability was answered. They had the first fully controllable glider and they made more than 1,000 flights with it above the sand dunes at Kitty Hawk, North Carolina.

So, in 1902, the Wrights returned to their home at Dayton, ready to develop a powered version of their flying machine. This kept many features of the glider – the biplane design, the front elevator and rear rudder, for example. But it was much larger (its forty-foot [12.3-metre] wingspan was almost twice the size of the glider's). And, of course, it had an engine, a four-cylinder internal-combustion engine placed in the middle of the lower wing and turning two pusher propellers. Take-off was to be from a dolly on a single-rail track, and it had a pair of skids on which to land.

It was towards the end of 1903 before the Wrights were ready to try out *Flyer* 1, as the new machine was called. The test flight was set for 14 December. The brothers tossed a coin to decide who should adopt the pilot's prone position next to the engine for the first flight. Wilbur was the winner. He set off along the take-off rail, but made a mistake in his setting of the elevator and the aircraft hit the sand at the end of the rail and came to a grinding halt.

Repairs were necessary, and they had to wait until December 17 to try again. Now it was Orville's turn. He achieved take-off, and made a successful twelve-second flight. The brothers, alternating turns as pilot, made three further flights, the fourth lasting for fifty-nine seconds and covering 852ft (260m) of powered, controlled, human-carrying flight. By the following year they were making successful turns in the air, and by 1905 they had made two improved aircraft; fuel capacity seemed to be the only limitation on the amount of time they could stay in the air. They had made a human dream come true, and created an invention that would change the face of transport, and, effectively, reduce the size of the globe.

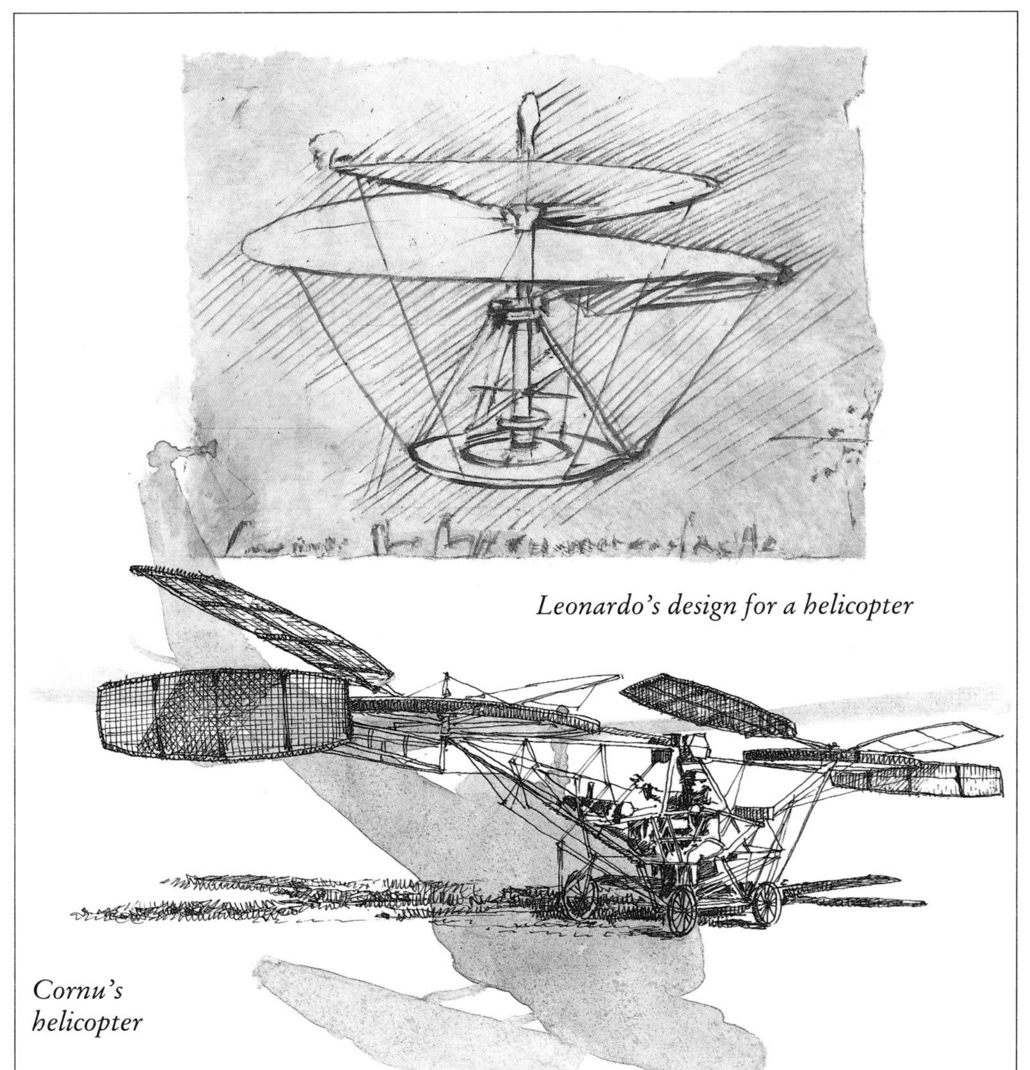

Leonardo's design for a helicopter

Cornu's helicopter

The helicopter

From early on the idea of an aircraft given lift by some sort of spinning rotors or discs seemed attractive. Such a machine would, it was hoped, be able to take off vertically, removing the need for the catapults and rails that dogged so many early aircraft designs. Leonardo da Vinci sketched a helicopter in the early sixteenth century. Aircraft and glider pioneer George Cayley also came up with a design in 1843. But a working, human-carrying helicopter did not appear until the beginning of the twentieth century. One design that had great promise was tested at Douai in 1907. It was built by Louis and Jacques Breguet, and Professor Richet, and had four large rotors arranged around the pilot, who sat at the centre of the machine. It was only ever tested while being held down or tethered, but achieved a lift of five feet (one-and-a half metres).

The first, free, piloted flight occurred during the same year, in a machine both designed and piloted by a French mechanic, Paul Cornu. The flights he made were brief and low, but his twin-rotor helicopter showed enormous potential and he abandoned it only because of a lack of money. Had he continued, he would probably have had persistent problems of control. These were not solved until Juan de la Cierva invented the autogiro, with its independently articulated rotor blades, in the early 1920s. During the following twenty years the modern helicopter was developed, with the American engineer Igor Sikorsky playing a leading role.

SCRIPTS AND ALPHABETS

Originally devised for keeping records of trade and accounts, writing was to prove invaluable in spreading ideas and information

For thousands of years people have felt the need to record information visually, in a way that other people can understand, or 'read'. It is possible that some of the first examples of this type of 'writing' may be the marks found on the walls of caves used by our ancestors of the neolithic period. Such marks include, as we have seen in the section on art (page 22), images of animals, stencilled handprints, and abstract patterns.

These are unlike writing in the modern sense, in that there is no recognizable code that we can interpret. Ideas are more effectively recorded by modifying or decorating objects, as examples from more recent cultures show. Australian aborigines used message sticks, made of wood and bearing marks or grooves. The Iroquois of North America used the patterns and colours of their wampum belts to transmit messages.

While these devices could carry quite detailed messages, the need was eventually felt for more sophisticated systems. This was particularly true in those societies that were just on the brink of producing a food surplus, and thus also beginning to trade. As they did this, they started to keep systematic records of goods traded, of possessions, and of taxes due and paid. It was therefore the need for account-keeping, and the desire to mark personal possessions that led to the first, true, written scripts.

Cuneiform

One of the first places where this happened was in Ancient Mesopotamia. By the fourth millennium BC, writing was well established here. To begin with, this was picture-writing: a stylized image of an object would stand for the object itself. The result was a script that was both complex (there were at least 2,000 signs) and cumbersome to use. So the signs gradually became more abstract, making the writing process, at least, more straightforward. Eventually, the pictographic system evolved into a form of writing that was totally abstract, made up of a series of wedge-shaped marks, and with a much smaller number of characters. This form of writing is known as cuneiform (from the Greek for wedge-shaped), and was written on wet clay tablets using a wooden stylus with a wedge-shaped tip. When the tablets set hard they provided an almost indestructible medium for information storage.

Pictograms were difficult to master and, although cuneiform was much easier, writing was at this time mainly the preserve of professional scribes. For the purposes of showing ownership a seal was therefore often used instead. Early seals might bear simply a design personal to the owner. Later on, a written inscription would also be included.

Sumerian clay tablet

Cuneiform was highly successful. Thousands of clay tablets have been unearthed bearing trading and taxation records from the Mesopotamian cities, and cuneiform script was used to write the Sumerian language. But cuneiform was also adapted to write the Babylonian and Assyrian languages, tongues quite different from Sumerian. Although cuneiform was much less well suited to these languages, the script was nevertheless used widely in the Middle East for a wide range of documents, from trading records to the letters of kings. The success of cuneiform was at least partly due to the fact that its wedge-shaped marks were very well suited to the prevalent writing medium – the clay tablet.

The scripts of Ancient Egypt

Cuneiform characters worked in several ways. Some were derived from pictograms and represented particular objects or something associated with an object. Some were also used to represent syllables – so it was possible for one sign to represent an object, or another object with a name that sounded the same in the Sumerian language, or a similar-sounding syllable in another word.

Egyptian hieroglyphs formed another successful script that used pictographic signs, but adapted for different purposes. This system was used for some 3,000 years by the Egyptians. During this time it retained its pictorial form: the hieroglyph for 'eye', for example, was a drawing of an eye; that for 'weep' was an eye with the addition of lines to represent tears. Such signs could also be used to represent syllables of the same sound. In addition, there were twenty-four signs representing single consonants, with which words could be made up if necessary.

Hieroglyphs were well suited to the art of the letter-cutter (the word hieroglyph means 'sacred carving') and were widely used for carvings on monuments and tomb walls. But they were difficult to write quickly, and Egyptian scribes

In the background a reed stylus is used to make wedge-shaped marks on a clay tablet, producing the cuneiform writing of the Sumerians. In the foreground are examples of various early scripts – Egyptian, Roman and Chinese.

ONE PALM
ONE HAND
ONE FIST
TWO PALMS
ONE ROYAL CUBIT 523·5mm

evolved another style, which was faster to write using a pen on papyrus. This script was a more stylized version of hieroglyphs and was known as 'hieratic'. Where fast writing was required, for example, in the spheres of administration, records and accounts, hieratic, or its later derivative, 'demotic', were used.

The coming of the alphabet

It is interesting that, with their twenty-four consonantal signs, the Egyptians had the basis of a perfectly workable alphabet. But they did not take the step towards a fully alphabetic system of writing. The origins of the Western alphabets lie not in hieroglyphs, but in the Semitic scripts of the peoples of the eastern Mediterranean. One important people from this area, in this respect, were the Phoenicians, who lived along the shores of Syria and Lebanon. From here, after about 1600 BC, they developed an extensive seafaring empire, trading and setting up ports all around the Mediterranean. They used a script made up of consonants, so it was not a true alphabet. Because they were such great traders, the Phoenicians left their inscriptions all over the area – in Cyprus, Sardinia, Malta, Sicily, Spain, at Marseilles, on the coast of North Africa, and in Greece.

The Greeks used many different scripts to write their language. Among these were the Cypriote script and Linear B, the latter script found in the palaces of the Minoan civilization on Crete. Both of these were syllabic scripts (that is, their signs represented whole syllables rather than individual letters), but neither was well suited to the Greek language. Nor was Phoenician, since vowels are highly important in Greek, and the Phoenicians had no way of writing them. But the Greeks learnt to adapt the script, using the consonantal signs that were little used in their own language to represent vowels. This was a vital step; once they had found a way of representing vowels, the Greeks had an efficient way of writing down their language – an alphabet (the word derives from alpha and beta, the first two letters of the Greek alphabet) had been created.

The alphabet the Greeks used did not remain static; they modified the script later to incorporate other sounds. The Greeks also changed the direction in which the script was written. At first, like its Semitic ancestors, it was written right-to-left. Next there was a period in which the script was written right-to-left and left-to-right on alternate lines (this method is called 'boustrophedon', a Greek word meaning 'in the way an ox ploughs a field'). Finally, the convention

Number systems

Various number systems have been used around the world, many of them evolving independently. The Ancient Egyptians and Babylonians used numbers based on multiple strokes, or wedge-shaped marks. The Maya of central America had a similar system that used dots. The Greeks and Romans adapted alphabetical symbols for use as numbers. The modern decimal system, with its vital zero symbol, was evolved from early Indian numbers.

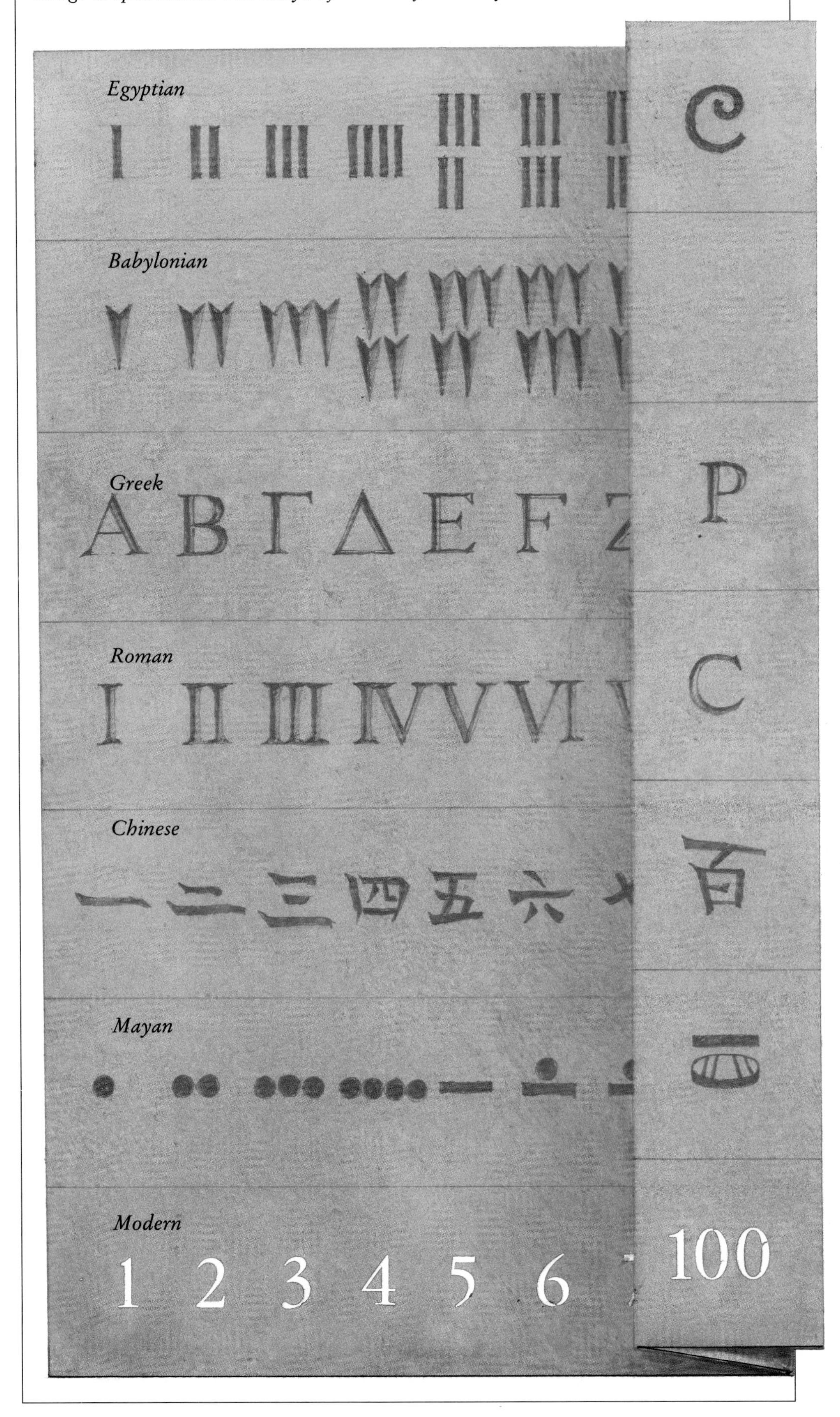

Evolution of an alphabet
There were many stages in the development of the Roman alphabet. These examples show how some of the letters of the Phoenician script evolved, through scripts such as Hebrew, Greek and Etruscan, to the Latin script used in the Roman Empire. This is virtually the same as the alphabet used so widely today, although it is adapted – for example, with the additions of accents or the omission of certain letters – for use with different languages.

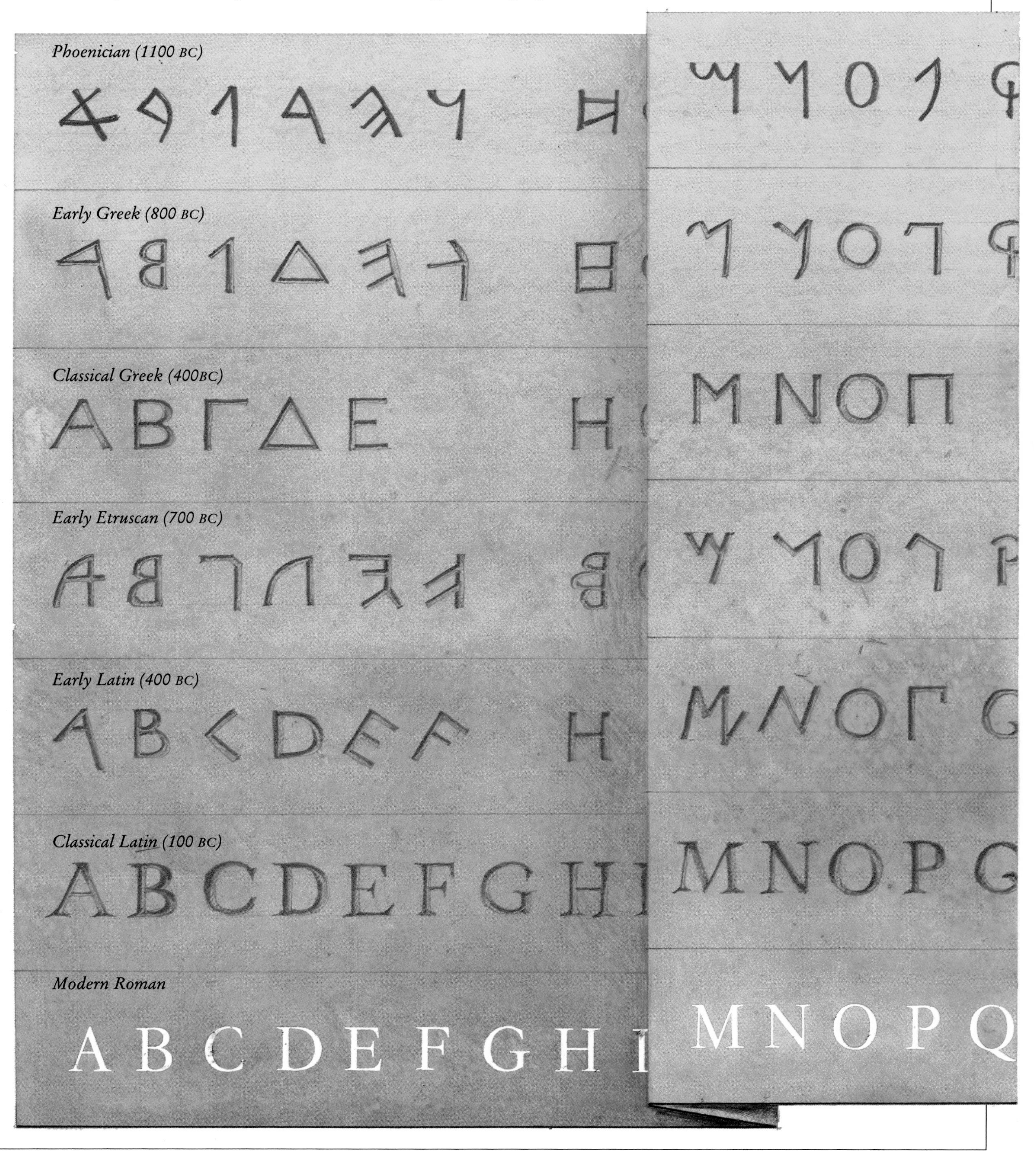

of left-to-right writing was established. At the same time, non-symmetrical letters changed their form, so that the loops on the letter 'B', for example, pointed towards the right, as they do now, rather than to the left.

The script changed in other ways, too. Different writing materials required different approaches. For example, many early examples of Greek were chiselled on stone. This approach was helped by having a script largely made up of straight lines. For writing on papyrus (or later, parchment) a more flowing, rounded style was developed.

The Roman alphabet used in so many places today was derived mainly from the Greek. Roman letters were very similar to the Greek ones, although some of the sharp angles of certain Greek letters were made more rounded in the Roman alphabet (Δ became D, for example, and Σ became S). In addition, some of the Greek letters were dropped from the Roman alphabet – Latin did not have sounds such as 'th' or 'ph'. However, in general, the Roman alphabet of the seventh century BC is the alphabet still used in Western countries today.

Arabic

Many early scripts, such as cuneiform and Phoenician, appeared, and were successful because of trade. They allowed merchants to keep records of deals and governments to write down what people owned and owed. A frequent and interesting by-product of this was the emergence of a fresh literary culture. The early epic poems of Mesopotamia, and the great literature of Ancient Greece owe a debt to the Sumerian and Phoenician traders who developed the early scripts.

Business was far from the only influence on writing. In the Islamic world, for example, the sudden emergence of the new religion of Islam stimulated the development and spread of Arabic script. This script had evolved, probably during the fourth or fifth centuries AD, from the writing of the Nabataean people, who lived around Sinai. It is based on the original signs of the Semitic languages, with the addition of further signs to indicate the subtleties of pronunciation.

The Islamic faith was founded in the seventh century. The text of the Koran was dictated by God to the Prophet Mohammed, and had to be copied by the followers of the faith. These words obviously have a key importance for the faithful, and are meant to be written down with accuracy and reverence. Consequently, the skill of writing and the art of calligraphy were, and are, highly regarded in the Islamic world. The fact that the representation of living forms is forbidden in Islamic art made calligraphy even more important. In addition, writing became vital as a way of recording Islamic learning and scientific discoveries.

Other non-alphabetic scripts

Just as the eastern Mediterranean was a crucial area for the development of writing in both the West and the Middle East, there was another script that proved influential in eastern Asia – Chinese. Here the development was very different. Whereas in the West the tendency has been for the number of signs to decrease and for the idea of the alphabet to develop, in China the script has remained non-alphabetic and has become more, not less, complex. In addition, the Chinese script has altered little in essence since it was first devised – it has not undergone the often radical transformation found in many scripts in the West.

Chinese signs represent concepts. They do this in a number of different ways. For example, some of the signs are stylized pictures of objects, some are combinations of these signs to indicate more complex of abstract ideas, some are combinations plus additional signs to indicate sound.

As will be clear from this, the number of different signs needed is vast – in the Shang period (*c*.1766–1122 BC) there were about 2,500; today there are some 50,000. This makes the script difficult to learn. However, it does have one important advantage which was especially useful in China: the script can be read independently of the spoken language. In China, a large country with a population speaking many different dialects but ruled from a central point, this feature proved invaluable. Furthermore, it meant that the script could be adopted in other countries, and as Chinese influence spread, so did their method of writing.

One place that felt the influence of

China very early on was Korea. Chinese Emperor Wu Di conquered Korea in AD 109 and many Chinese followed him there. They took their script with them, and this survived, even though the Chinese rulers left Korea less than half a century later. Chinese script remained in use there for centuries, although it had to be modified to suit Korean, a very different language from Chinese.

The Chinese influence was even stronger in Japan. There was a Japanese invasion of Korea in AD 370, and Korean experts in Chinese language and script soon arrived in Japan. The spread of Buddhism also brought much interaction between China and Japan, the latter country taking on board Chinese script and adapting it to the local language.

The Koreans and Japanese altered the Chinese script in different ways. The Koreans added extra characters to create a syllabic script; the Japanese used some Chinese signs syllabically and added phonetic symbols. Such adaptations show how the cultural domination of a people can foster the spread of their script, even in an area where it is not ideally suited to the language.

The encouragement of trade, the spread of political influence, the recording of literature and the transmission of religious ideas are but some of the most important results of the use of writing. Whether the script is alphabetic or based on some other principles, the effect on the availability and transfer of information has been far-reaching and lasting.

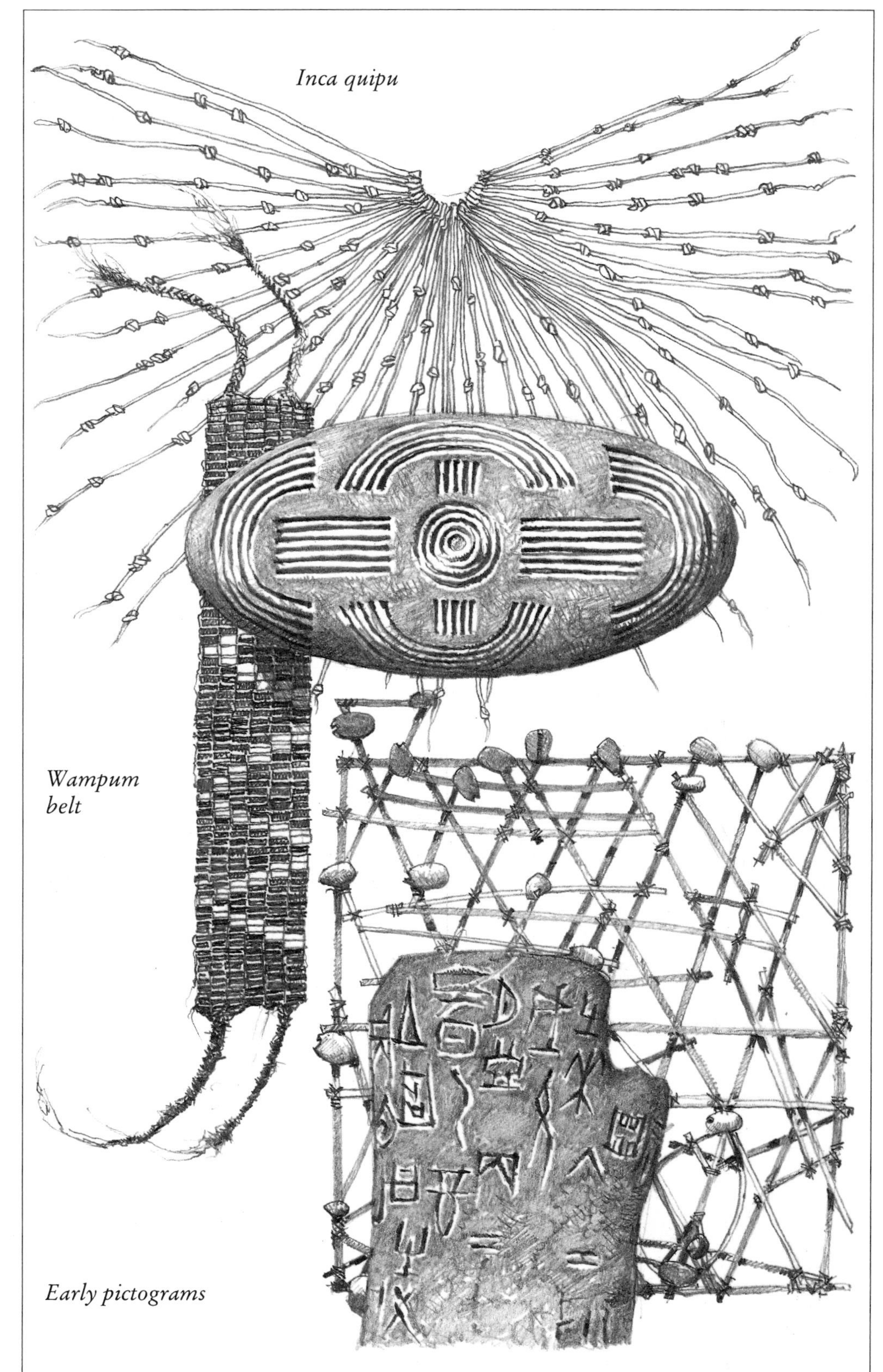

Forerunners

What we normally think of as a written script is not always necessary for recording information. Many societies have made good use of non-written recording media. A well-known example is the 'quipu' of the Incas, a device consisting of variously coloured and knotted cords, used to record information. Even in the West, centuries after written scripts had taken hold, similar systems were sometimes used. Tally sticks with cut notches, for example, recorded financial transactions in the medieval English exchequer. Yet more good examples of this type of information-recording are the patterned wampum belts of some North American Indian peoples, and the message sticks used by the Australian Aborigines.

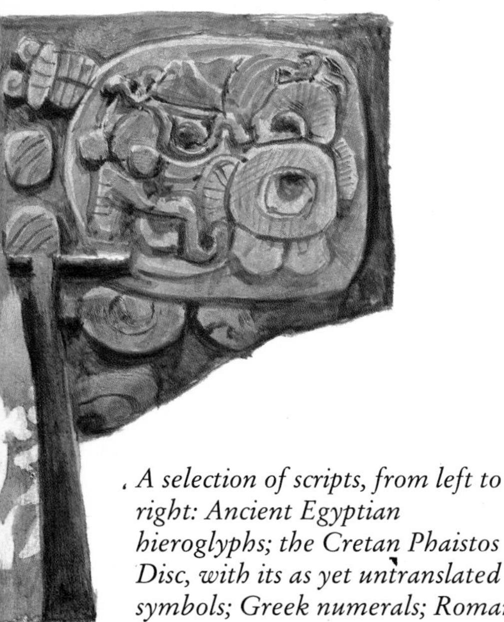

A selection of scripts, from left to right: Ancient Egyptian hieroglyphs; the Cretan Phaistos Disc, with its as yet untranslated symbols; Greek numerals; Roman letters; Arabic script; Chinese characters; and Maya glyphs

THE BOOK

Durable and legible means of transmitting knowledge to numerous people began with clay tablets and progressed to the printed page

The first writing was used for legal, administrative and accounting purposes, not the kinds of texts that normally come to mind when we think of books today. However, scribes soon began to realize the potential of cuneiform for preserving other sorts of writings – hymns used in the temple, spells and texts used to predict the future, and stories that had previously been handed down by word of mouth. Such texts, first written down in Mesopotamia before 2000 BC, form the beginnings of what we now think of as literature, and the origins of the book.

Clay tablets

If these texts, written in a dead language by unknown scribes, seem remote, we should remember that they contain the germs of so many books to come. Take the most ancient narrative poem, the *Epic of Gilgamesh*, unearthed on clay tablets in the library of the Assyrian King Assurbanipal (669–627 BC), but dating to a much earlier period. This story contains an account of the great flood, made famous in the Bible, foreshadows the myth of the labours of Hercules, and deals with such enduring subjects as the search for immortality, and relations between man and woman, and man and man.

Assyrian King Assurbanipal

Important as all these themes are, it is not the purpose of this book to trace their course through the literature of the ensuing centuries. What we are concerned with here is rather the medium in which the ideas have been transmitted. The clay tablet, on which early books were written, had several advantages. The material was freely available. When it had been dried in the sun, it was hard and durable. The cuneiform script was well adapted to be written on damp clay.

But the clay tablet also had its limitations: it was heavy and cumbersome; it took a lot of space to store, and was not as easily portable as subsequent media. It was thus a better medium for recording and storing government accounts (kings had plenty of storage space) than for letters or literary texts.

Papyrus

In Ancient Egypt, the hieroglyphic script was well suited to cutting inscriptions on memorial tablets and tomb walls. But in a civilization of the complexity of Egypt it is not surprising that a more portable writing medium was also developed, in which the Egyptians made use of the papyrus reed that grew in abundance by the River Nile.

The reed has triangular stems with a green outer skin and white pith inside. To make a material for writing on, the Egyptians first removed the outer skin. They then cut the pith into thin strips, and laid these strips together with their edges overlapping. When the required size was achieved, they laid another layer of strips at right-angles to the first. A heavy weight was put on top of the two layers, and they were left to dry and fuse together.

The result was a thin, flexible, writing material. Papyrus was widely used in Ancient Egypt for recording everything, from tax records to the Book of the Dead.

Chinese woodblock and impression

It was well suited to use with the reed brush, and the Egyptians evolved scripts that were ideal for this type of writing. Papyrus continued in use for thousands of years, not only in Egypt, but in later civilizations such as Greece and Rome. Many of the books that have come down to us from the early western civilizations, therefore, were written on papyrus, the material having survived thanks to the arid climate of Egypt and the eastern Mediterranean.

Parchment

Much of this papyrus continued to come from the banks of the Nile. There could therefore be problems when the supply was interrupted, such as happened in the second century BC, when Egypt stopped supplying Pergamon, in Asia Minor, with papyrus. The story is that the scribes at Pergamon turned to a new material, leather, so beginning the vogue for parchment. In fact, parchment had been used long before, but it became more and more common during the Dark Ages and then the Middle Ages.

Parchment was made from the skin of sheep, calf or goat. The skin was soaked in a lime bath and then thoroughly cleaned.

It then had to be burnished or scraped with a pumice stone or knife before it was ready for use. The majority of the European books of the Middle Ages were written on parchment with a quill pen, in one of the 'scriptoria' attached to nearly every monastery. The beautiful illuminated manuscripts so typical of the Middle Ages evolved because parchment and the goose quill were so well matched. Sheets of parchment could also be sewn together easily into a volume, so that the pages were protected between stout covers of leather or wood. Books began to take on their familiar bound form for the first time.

In the West, the bound parchment volume remained the typical book for hundreds of years. The situation was different in Asia, where materials such as strips of bamboo or rolls of silk were used for books. Another development in Asia looked towards the future: this was printing. The first printed books were made by carving text or pictures on wooden blocks, which could be inked and used to stamp an impression. Such books began to appear in China in the ninth century AD. Individual pieces of type that could be re-used were also developed in China in the eleventh century, and in Korea in the thirteenth and fourteenth centuries. But printing was to find its greatest development in Europe, where the Roman alphabet, with a mere twenty-six letters, proved a more obvious candidate for movable type than Chinese, with its thousands of different characters.

Printing – some key technologies

Several different technologies had to be in place before printing could be carried out efficiently. First of all, a suitable material was required on which to print – paper had to be available in sufficient quantities. Next, an ink of the right consistency and colour was needed. Then there was the press itself. Finally, movable type was needed to print the letters.

In some ways, the press was the most straightforward of these items. Presses had already existed in Europe for centuries, not originally for printing but for pressing ingredients such as apples for cider or liqueurs.

Papermaking

Paper had to wait longer to arrive on the European scene. The earliest paper comes from China. The oldest piece that has survived is made of hemp fibres and was found in Shensi Province; it dates from between 140 and 87 BC. It is thick and uneven and unpromising as a writing material: in any case, the Ancient Chinese preferred bamboo or silk to write on. However, there exists some paper dating from about AD 110 that does bear writing, and paper was probably used more as a writing material in China as the manufacturers learned to produce a finer, more consistent product. This paper was usually made from linen rags, pounded in water to make a pulp. A tray containing a wire or bamboo grid was lowered into a vat of the pulp; the grid was then removed, together with a thin sheet of pulp. When the surplus water had been removed, the sheet was taken off the grid and left on a piece of felt. Later, it was hung up to dry.

Medieval illuminated initial

The industry probably arrived in the West by more than one route. For example, a Frenchman who had been imprisoned at Damascus set up a mill in 1157; meanwhile, the art of papermaking was also transmitted westwards through Muslim Spain. By the following century, there was a paper industry in Italy (with a mill working at Fabriano, for example, by 1276). This industry had been well developed in the East before it came to Europe.

From kites to clothes, wallpaper to toilet paper, this material played a significant part in Chinese life during the Middle Ages. However, it did not travel westwards until the seventh century, first to India and then, about a hundred years later, to the Arab world. The Arabs developed a large papermaking industry, which spread from Samarkand, to cities such as Baghdad, Damascus and Cairo. From here, paper was exported to the West, but only in the twelfth century did papermaking begin in western Europe.

The increasing availability of paper itself created a stimulus in the book market. Baghdad was an important centre; in AD 900 the city had around one hundred workshops where scribes and binders were copying and putting together books for sale. Centres of papermaking were also often established near centres of learning. Toledo in Spain, and Fez, with its Islamic university, in Morocco, are two such examples.

Ink

Inks for writing had been made in many different ways. Juices from fruits and vegetables, sepia (the liquid obtained from the ink-sacs of cuttle fish), and suspensions in water of a variety of substances were used. The earliest surviving inks come from Ancient Egypt, where a solution of lampblack (a pigment made from soot) in vegetable gum was used to write on papyri. Later, water-based inks were more common.

Some of these could be used for printing with a woodblock, but with metal type there were different requirements. The metal did not hold water-based inks adequately, so an oil-based ink had to be used. Lampblack in a glue-like base of linseed oil provided the answer. Such an ink worked well on the press and produced a good black image.

Woodblock books

By the late Middle Ages, then, paper, ink and press were all available. It is not surprising, therefore, that a thriving printing industry had begun in Europe, and it began in the place where hand-made books were already being produced: the scriptoria of monasteries. To begin with, it was a very simple process. Small stamps were created to hand-print regularly appearing decorative elements in manuscript books.

It was quickly realized in the religious orders that such woodblocks could be useful in producing large numbers of religious images for wide consumption. The result was a trade in woodblock prints of religious subjects such as the Virgin and 'popular' saints, with which people could decorate their homes. At first these had no text, but the carving of short religious messages was not too arduous for the block-maker, and increased the effectiveness of the image, so words and pictures began to be printed together.

The next step was short booklets, often with quite extensive text, but still highly illustrated, and printed by the woodblock process. The words were usually in the vernacular, opening up a wider readership than the Latin texts of the illuminated manuscripts. And so the effectiveness of printing as a tool for spreading ideas, information and propaganda began to be appreciated.

Simple woodblock books like these were highly effective. They could reach a wide audience, and they had the advantage that they were very easy to produce. A minimum of equipment (blocks of wood, a knife, a small, simple press) was required. They caught on, and were being produced long after more sophisticated methods of printing had been introduced.

The coming of movable type

Woodblocks provided a good method of printing so long as the text was not too long. But printers must have puzzled long and hard about ways to overcome the enormous waste of time and effort involved in printing words with a woodblock. Once the laborious process of carving had taken place and the book been printed, the block was useless. The great

The first printed books
Once type had been cast, it was placed in a case like the one shown on the right. The letters were then arranged in lines and pages, and several pages at a time were put on the press. The basic processes of casting, setting and printing are described opposite. The result was books that were designed to look like the manuscripts that preceded them. They had Gothic type, and decorated initials were added by hand. Illustrations could be printed, too, from wood blocks.

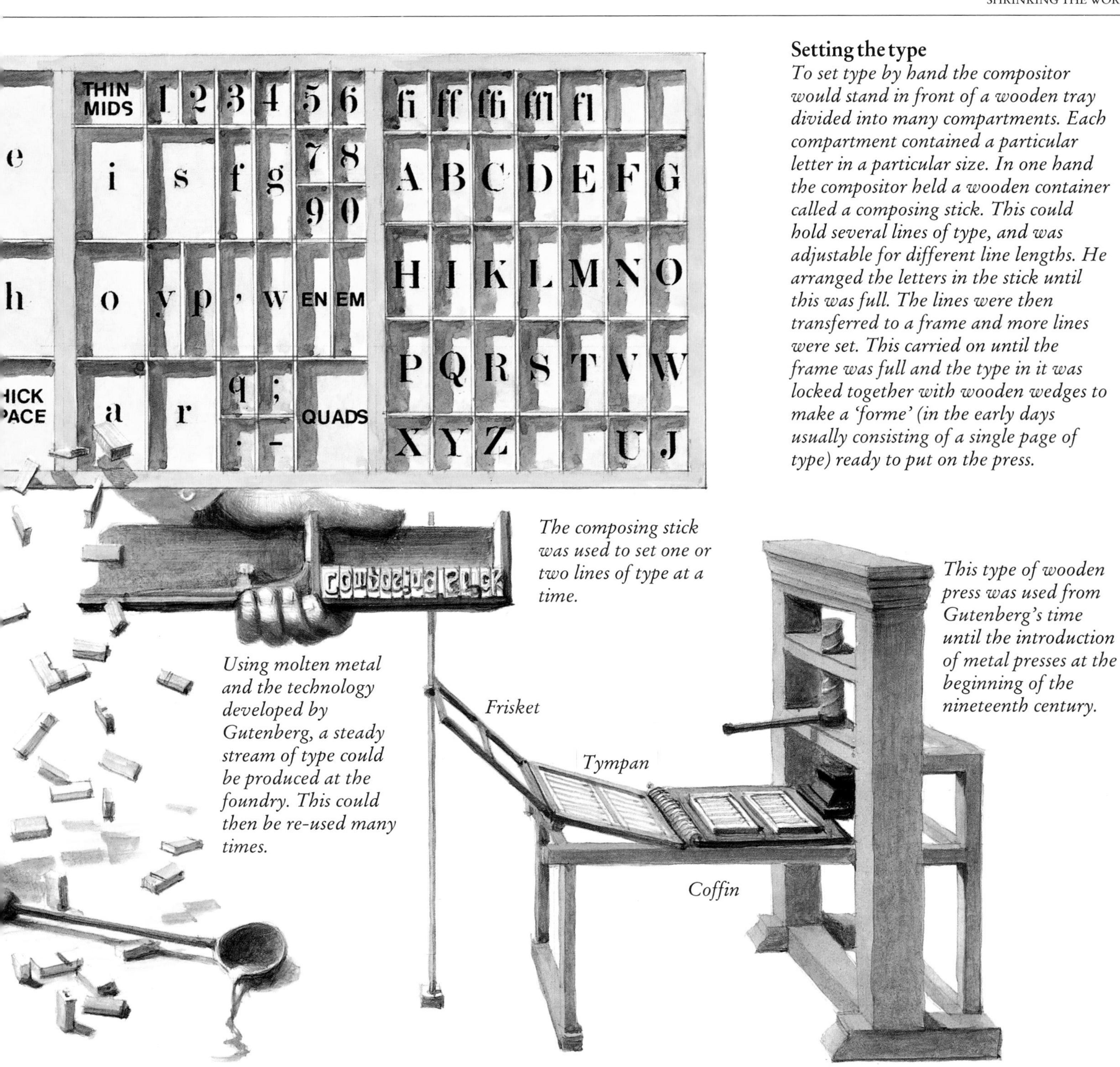

The composing stick was used to set one or two lines of type at a time.

This type of wooden press was used from Gutenberg's time until the introduction of metal presses at the beginning of the nineteenth century.

Using molten metal and the technology developed by Gutenberg, a steady stream of type could be produced at the foundry. This could then be re-used many times.

Setting the type

To set type by hand the compositor would stand in front of a wooden tray divided into many compartments. Each compartment contained a particular letter in a particular size. In one hand the compositor held a wooden container called a composing stick. This could hold several lines of type, and was adjustable for different line lengths. He arranged the letters in the stick until this was full. The lines were then transferred to a frame and more lines were set. This carried on until the frame was full and the type in it was locked together with wooden wedges to make a 'forme' (in the early days usually consisting of a single page of type) ready to put on the press.

Casting the type

The first step in the process was to create a metal punch with a positive image of the letter to be cast. This was made by hand in a hard metal. A piece of softer metal was then struck with the punch to make a negative matrix. Any number of matrices could be made from the same punch. The matrix was then placed inside a mould and the mould closed. Next, the type-caster would pour hot metal into the mould and allow it to set solid. When the mould was opened, a piece of type fell out. The piece of type would have a rough end where the metal was poured into the mould, so this would be removed and the edge smoothed with a type plane. This would help to ensure that all the pieces of type were the same height – essential if the letters were to print and to print evenly. The entire casting process would have to be repeated many times to create enough pieces of type for each letter of the alphabet, including both upper and lower case letters, punctuation marks, and symbols. It was very laborious, but once the complete collection of pieces had been made the type could be used many times.

Working the press

The printer would place the forme on the carriage of the press and ink it with a pair of dabbers. Then the paper was put into the tympan and the protective frisket folded over to mask the areas of the sheet that were not to be printed on; the tympan was folded in turn over the forme so that the paper was in position above the type. The printer would now push the coffin under the platen. This could then be screwed down to make the impression. The process seems slow and cumbersome but early printers managed to print up to three sheets a minute using this method.

breakthrough would be finding a way of making individual letters quickly in large numbers – and if possible making them in a way that allowed them to be re-used.

The method that was evolved to do this was completely unlike anything to do with the woodblock process. The people who founded the modern printing industry were not monks or woodcarvers but metalworkers. In a way, this was not surprising. Wood was not an easy material to use to print letters. It was difficult to cut the letters cleanly enough; the wood could warp, destroying the evenness of the print or making the block useless; and the wood wore out quickly. Towards the middle of the fifteenth century, a number of goldsmiths, working independently in Mainz, Basel, Avignon and elsewhere, began to look for ways of improving the process by using metal instead of wood. Similar work was carried out independently in Korea about fifty years earlier.

The most well known of these pioneers was Johann Gensfleisch, known as Gutenberg. A citizen of Mainz who settled in Strasbourg, Gutenberg set up in partnership with three other men with a view to developing some secret processes that Gutenberg had invented. One of these processes was the polishing of precious stones, another involved mirrors, the others were to do with printing.

We know a little about these processes from documents of the time, but unfortunately the documents are difficult to interpret. They are written in rather vague terms, and the technical words are not defined, probably partly because the techniques they described were still being developed, partly because Gutenberg was trying to keep his work secret. It is difficult to tell what is meant in the document by 'pieces' or 'formes', for example. However, it seems to be the case that what Gutenberg and his partners developed was a method of type-casting, a reliable way to produce re-usable letters with relative ease and speed.

Gutenberg and his partners were not the only people developing the technology of printing in this direction. Procopius Waldvogel was another goldsmith, this time from Prague. He made agreements in the 1440s to teach people in Avignon the skills of 'artificial writing'. 'Alphabets of steel' and other metal equipment mentioned in documents about Waldvogel suggest that he too was developing some sort of metal type. Other metalworkers in Holland and Belgium also appear to have been developing movable type at around the same time. But we cannot be sure exactly what all these people were doing.

What is clear, is that by the 1450s books were being printed with movable type. Gutenberg, in partnership with Johann Fust (who provided the money) and Peter Schoeffer (a calligrapher turned printer), had a press at Mainz, which produced its first datable product, a Psalter, in 1457. By this time, it seems that Gutenberg himself had left the business, possibly elbowed out by Schoeffer and Fust, under whose imprint the Psalter appeared. By this time the famous forty-two line Bible (it has forty-two lines per column on the page), traditionally known as the Gutenberg Bible, and often referred to as the first printed book, had been printed.

These early volumes were the start of a revolution in communications. If Gutenberg had tried to keep his inventions secret, once he started to print, news got out. Soon printshops were being set up all over Europe. By the end of the century

Close-up of early Gothic type

many cities had printing presses. Augsburg, Basle, Bologna, Cologne, Deventer, Florence, Leipzig, Lyon, Milan, Nuremberg, Paris, Rome and Venice all had a big enough printing industry to produce over 500 volumes before 1500. And these cities were only the major centres. Presses were springing up everywhere, with astonishing rapidity for the times. It is now thought that almost 40,000 titles were printed in Europe before the end of the fifteenth century.

Suddenly, books were cheaper and more widely available than had been imaginable a few decades before. And many of the newly printed works were, like the block books that preceded them, in the vernacular. Ideas were no longer the preserve only of those who could read Latin. At first, the demand was for religious works – Bibles, psalters, prayer books, and so on. However, soon there came books of law, scholarship and science, together with translations from the classics of Ancient Greece and Rome. Both the culture of the Renaissance, and the ferment of ideas created by the Protestant Reformation owe a great deal to the spread of printing.

Improvements in technique

We do not know exactly how long it took to bring the printing press and its movable type to the level of efficiency that allowed rapid printing of large numbers of books. There certainly must have been many problems with the original equipment. Having cast the type, it would have proved difficult to keep a whole page's-worth of the tiny letters securely together while the actual printing took place. There is some evidence that early printers experimented with making a hole in each piece of type, presumably to allow some sort of metal rod to be pushed through an entire line of type to help keep it in position. This must have proved very tedious, and printers eventually settled on a method of keeping the type in the page-size 'forme' with simple wooden wedges.

There were further problems associated with the inking of the type. The type was inked with a pair of hand-held circular 'dabbers'. This was difficult when the type was inside the press, so the bed on which the type lay was mounted on a sliding carriage. This could be pulled out to allow the type to be inked, and pushed back in again when the printer was ready to make the impression.

Often the ink did not stick evenly to the paper, or got on to the margins of the sheet. The first problem was solved by putting several sheets of paper underneath the sheet to be printed. This provided sufficient play to allow any characters a fraction lower than the majority to print. The second difficulty was overcome with a rectangular mask called a 'frisket', which obscured all the non-printing areas on the paper when the impression was made.

Numerous other improvements were made to the printing process, including the progressive use of more and more metal parts in the press. These included a spring or 'yoke' to give a still more even impression, a metal screw-thread, and various other metal parts to reinforce the structure of the press. By the beginning of the nineteenth century, presses were being made completely of metal.

Printing with movable type is one of the most important of human inventions. In creating the book as we know it today, it represented not only a radically new technology but a different, more efficient way of recording and transmitting ideas. It not only changed the world but enabled others to go on changing it.

JOHANN GUTENBERG

OPTICAL INVENTIONS

Experiments with glass led to ways of correcting defective vision and allowed scientists to observe the heavens and the 'invisible' world of micro-organisms

Glass was first produced in Egypt, before 2000 BC. This early glass was rarely transparent because of impurities in the ingredients, but, later on, when clear glass started to be produced, its tendency to distort an image must have been noticed quickly. Early glass-makers found it difficult to make glass with a flat surface, and it would not have been long before people noticed that different curvatures produced different effects when objects were viewed through them.

Crude lenses dating from about 2000 BC have been found in Crete and western Asia, but it was not until much later that lenses of any quality were made. Improvements in lenses were no doubt reflected in the interest in optics shown by the writers of Ancient Greece and Rome. Euclid, the Greek scientist of the third century BC, for example, wrote about the reflection and refraction of light.

However, it was probably the scientists of the medieval period who first publicized the usefulness of lenses. Some of their knowledge came from the Islamic world. The eleventh-century Arab writer Alhazen described the magnification one could achieve with lenses, as did the English writer Robert Grosseteste (c.1170–1253). Grosseteste explored the fields of vision, mirrors and lenses. He also made a 'burning glass' from a spherical flask filled with water. This medieval Englishman was showing how the lens could open up worlds that were as yet only indistinct – and hinting at a time when people would be able to see things invisible to the naked eye. For this they would need two later optical inventions, the microscope and the telescope, inventions that were implied in Grosseteste's writings when he suggested that combinations of lenses might be used for examining very small and very distant objects. Meanwhile, the great usefulness of lenses in correcting sight defects was obvious, and spectacles were not uncommon by

Antonie van Leeuwenhoek

the late Middle Ages. The greatest centre of spectacle-making was in Holland, and it was here that many of the important advances in optics took place.

The microscope

In the late fifteenth and early sixteenth centuries the Dutch spectacle-makers began to develop microscopes. These were simple devices, with a single lens of the highest quality available, and a screw thread to focus the lens on the object. One of the greatest pioneers of the microscope was the Dutchman Antonie van Leeuwenhoek. During his long life (1632–1723) he taught himself how to grind his own lenses, and was able to obtain magnifications of up to 150. He put his microscopes to good use, making many observations of specimens from the natural world, and discovering bacteria in the process.

Simple microscopes continued to be produced in the eighteenth century, although improvements in design changed their appearance. The screw barrel was introduced to focus the lens, and better mounts were developed for examining different objects.

Meanwhile, another type of microscope, the compound microscope, was undergoing a parallel development. This type uses two or more lenses to enlarge the image. Again, it probably appeared first of all in Holland. In general, even greater magnifications were possible with this type, although there was still some sacrifice of clarity. It is not known for certain who came up with the idea of the compound microscope. Various Dutch scientists, including Hans Lippershey of Middelburg, and Hans and Zacharias Jansen are all possible contenders. All these scientists were working during the late sixteenth and early seventeenth centuries. Shortly after this, compound microscopes were being made in Italy, England and the Netherlands.

Leeuwenhoek had an English counterpart in Robert Hooke. Hooke was a multi-talented scientist – the microscope was only one of his interests. He developed the compound microscope, and his observations with it were published in an illustrated volume called *Micrographia* in 1665.

Microscopes did not change greatly until about 1830. This date marked the appearance of lenses that were free from

Laid out on the bench are some of the key early optical inventions. Antonie van Leeuwenhoek's microscope (1) *looked unlike any modern instrument. It had a double convex lens mounted between brass plates. The compound microscope* (2) *looks much more familiar. Its two lenses were held in an adjustable tube. This example was made by Campani. Robert Hooke's microscope* (3) *used a lens to focus light on the subject. In the background, a writer takes advantage of an optical invention that predates all these instruments, and which is still in use today, a pair of spectacles, one of many pairs illlustrated in the manual seen at his right shoulder.*

1
2
3

Galileo and his telescope

The famous Italian scientist was the first to make important observations using a telescope. His instruments were refracting telescopes in long, leather-covered tubes. Such telescopes did much to improve people's view of the heavens, but they also posed problems. Large magnifications required long telescopes, which were difficult to keep still. There were also problems with colour distortion. The reflecting telescope offered a clearer image and was easier to use. Even so, later astronomers realized that an important adjunct to any telescope was a stand that would hold it firmly aimed at whichever part of the sky you wanted to observe.

Lens-making

The first lensmakers probably stumbled upon their product by accident among the small bulbous pieces of glass they were making for other purposes. To reproduce the effect they cast discs of glass in a mould and then ground and polished the glass by hand to get as near as possible to the required surface curvature. Later lensmakers used a lathe to grind the glass. One such craftsman was Ippolito Francini, who worked for the Medici family in the first half of the seventeenth century. Still later, it was realized that even better results could be obtained by cutting the lens out of a solid block of glass, rather than first casting the basic shape in a mould. The pioneer of this technique was Giuseppi Campani, who worked in Rome in the late seventeenth and early eighteenth centuries.

the colour distortion that made many early microscope images indistinct. By this time, the importance of the microscope in opening up undreamed-of worlds to human vision was clear to all.

The telescope

As with the other optical inventions, the origin of the telescope is not known for certain. However, the story is that Hans Lippershey, the same man who developed the compound microscope, held a pair of lenses up to the sky and noticed the magnified image; he had created the first telescope. Lippershey applied for a patent on the telescope in 1608. He also made a number of telescopes and sold them to the government of the United Netherlands for possible military use.

When he applied for his patent, Lippershey knew that several people already knew how to make telescopes. He had rivals in Middelburg, and others as far afield as France, Germany, England and Italy. But the most influential person connected with the telescope in its early days was the Italian scientist Galileo, who made one for himself in 1609 after hearing of the Dutch developments. He immediately recognized the potential of the instrument for observing the heavens, and spent the following years building still better telescopes and looking at the skies. One of his first discoveries came from his observations of the sun. He watched the different sunspots on its surface and, by keeping a record of the spots, came to the conclusion that our nearest star rotates on its axis every twenty-seven days.

Galileo went on to use his telescopes to make further observations about the universe, recording the phases of Venus, moons orbiting the planet Jupiter, and other phenomena. His observations eventually led to the breakdown of the then prevailing view of the solar system (which saw the Earth at the centre with everything revolving around it) and led Galileo into serious trouble with the church authorities.

Galileo was forced to recant his heretical views in 1633, but his work with his telescopes has come down to us to emphasize the importance of scientific observation. His successor Scheiner (his opponent from the Inquisition) would use darkened glass to look at sunspots, and it has been recognized ever since that looking at the sun through a plain telescope is highly dangerous.

The seventeenth century proved a fruitful time for the development of the telescope. Johannes Hevelius of Danzig produced very long high-power telescopes in the 1640s, and used them to produce a detailed series of maps of the Moon and a catalogue of some 1,500 stars. He was followed by Dutch scientist Christiaan Huygens, whose telescopes were still longer, and better made. Some of these instruments were very long, and needed strong stands if they were to be aimed accurately and trained continuously on the same spot in the sky. By this time lens grinders had designed specialist lathes to make surfaces to more precise specifications than was previously possible.

However, even with these advances, astronomers still had to put up with colour distortion and blurring. Isaac Newton, whose work on light and optics well qualified him, tackled this problem. He created a reflecting telescope, which contained a spherical mirror. The mirror concentrated the image at the eyepiece and the reflecting telescope eliminated colour distortion.

More recently, great advances in astronomy have been made with radio and X-ray telescopes. These detect invisible radiation from objects in space, focusing the signals on an antenna, amplifying them, and recording them. In this way, information has been amassed about heavenly bodies much too far away to be seen with an optical telescope.

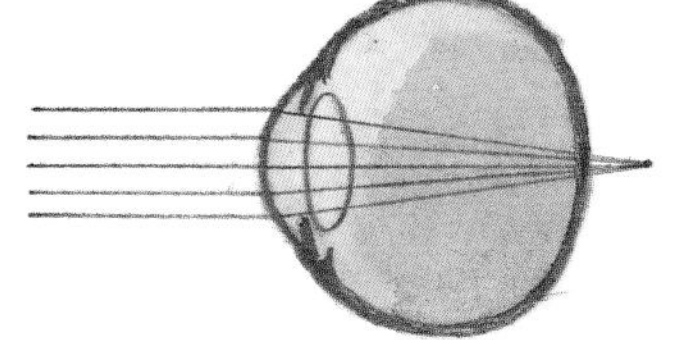

Long sight

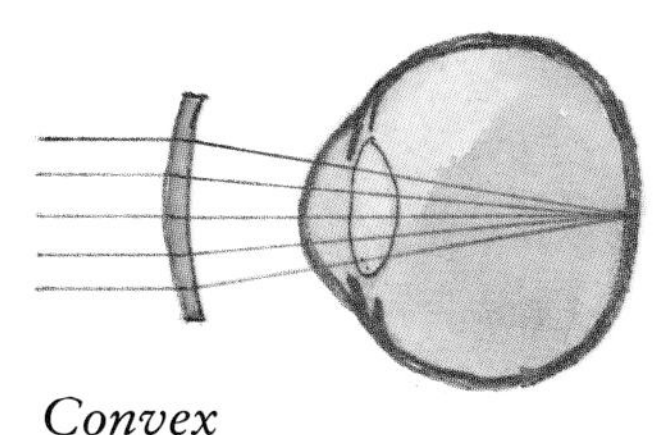

Convex lens *Long sight corrected*

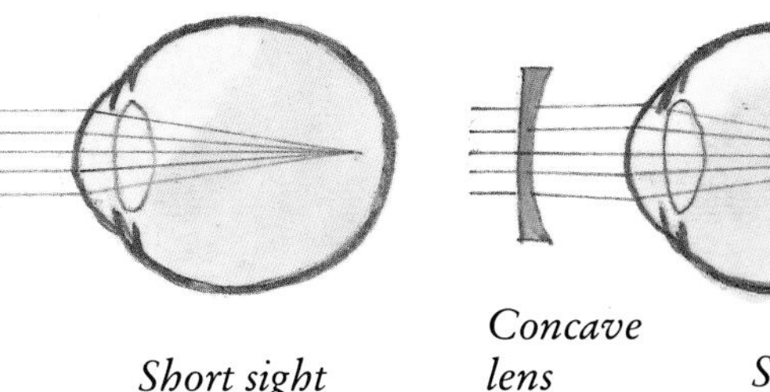

Short sight *Concave lens* *Short sight corrected*

Spectacles
The Roman Emperor Nero is said to have improved his vision by holding a jewel with curved surfaces to one eye. Since he did this at the arena, it may have been that the jewel had concave facets and helped to correct short sight. But such early examples of sight-improvement are rare. Spectacles with glass lenses first began to appear in the Middle Ages, probably in the late thirteenth century. These generally had convex lenses and were used to correct long sight. Concave lenses for the correction of short sight are first recorded in 1568.

REMOTE COMMUNICATIONS

Human ingenuity and progress are clearly shown in the methods that have been devised to send messages over great distances without endangering life or security

People have been trying to send messages over long distances for thousands of years. The Ancient Chinese, for example, lit bonfires along the Great Wall to warn of the arrival of attacks from the barbarian north. Native Americans were still using smoke signals as recently as the nineteenth century. The Romans flashed messages from hill to hill with mirrors. In later centuries, flashing lights and moving semaphore arms were used in similar ways. Methods such as these relied on sender and recipient understanding a particular code; they were also dependent on geography (signaller and recipient alike had to have good vantage points) and the weather; and a signal could be sent only as far as the eye could see (although you could achieve greater distances by using chains of signallers). The potential for a signalling system that did not have these drawbacks would be enormous.

Telegraph systems

It was not until the nineteenth century, with knowledge of electricity, that such signalling systems began to be developed. The first researchers into electricity noticed quickly that there was a relationship between electricity and magnetism. The relevance of this for signalling was that an electric current could deflect a magnetic needle. The great mathematician Karl Friedrich Gauss saw that this phenomenon could be used for signalling, and produced a simple experimental device in 1832.

The idea was taken up by British inventors W. F. Cooke and Charles Wheatstone. Their telegraph, which they patented in 1837, had five magnetic needles pointing at different letters. By using a code, they were able to reduce the number of needles to two.

Another important pioneer was the American Samuel Morse. He is remembered for his code of dots and dashes, which made signalling both by flashing lights and by telegraph more straightforward. An interesting feature of the code is its ease of use – it is simple to learn and the most frequently used letters can be sent the most quickly. In addition, Morse introduced relays to amplify the signal en route, thus extending the range and usefulness of the telegraph.

Samuel Morse

The telegraph fulfilled a need, and its use spread quickly. By the 1860s the East and West Coasts of the USA were connected by telegraph wires, and America and Europe by a transatlantic cable.

Telephony

But how much more useful would be a system by which the actual human voice could be transmitted. In this field the important pioneers were the Scottish-born (but USA-resident) inventor Alexander Graham Bell whose work in speech therapy led him to speech-reproduction devices, and the American Elisha Gray. Both of these men were developing their respective systems in the 1870s and both men filed patents on the same day in 1876. In Bell's transmitter, a piece of iron in a magnetic field vibrated when you spoke into it. A voltage was thus induced in the coils on the magnet and corresponding currents passed down the wire connecting from the coils to the receiver. Gray created a variable-resistance transmitter, which provided a more powerful way of transmitting speech.

Initially there was heated competition between Bell's company and the Western Union Telegraph Company, who used Gray's system developed by Thomas Edison. Eventually, Western sold out to Bell. Bell's company was well established by 1878, and Bell was already confronting the problem of how to link up all the telephone lines. The result was the concept of the telephone exchange, a switching centre to which all the telephone subscribers in a particular locality would be connected. It would then be possible for any subscriber to ask to be connected to any other subscriber in the area.

To begin with this was a purely local service, but it was soon realized that further exchanges could be introduced to link different towns and cities. And so, very quickly, the notion of a national telephone network began to emerge. Because different telephone companies using slightly different systems were involved, this idea did not take hold immediately everywhere. But in the USA, with the domination of Bell's company, and the pressing need for communications over long distances, the network spread rapidly. By 1885 there were already 140,000 subscribers and 800 exchanges. In other countries the growth was less spectacular, although in many nations it was not uncommon to see numbers of subscribers doubling every two to three years during the last years of the century. And, as the number of users grew, so more and more of the world's cities were interconnected with a growing network of telephone wires.

Alexander Graham Bell uses early telephone equipment. Behind are some of Bell's rough sketches showing the principle of his telephone.

A
B
C
D
E
Semaphore code
Semaphore, developed in 1794, used moving arms worked by ropes to create symbols for each letter.
Morse code
Samuel Morse's code of short and long sounds was ideal for transmitting along a wire with a 'key' (far right). The code is shown here both as dots and dashes, and as an electrical signal.
Five-unit code
This is a development of Morse code and was designed for use with a teleprinter. Each symbol is made up of five units.

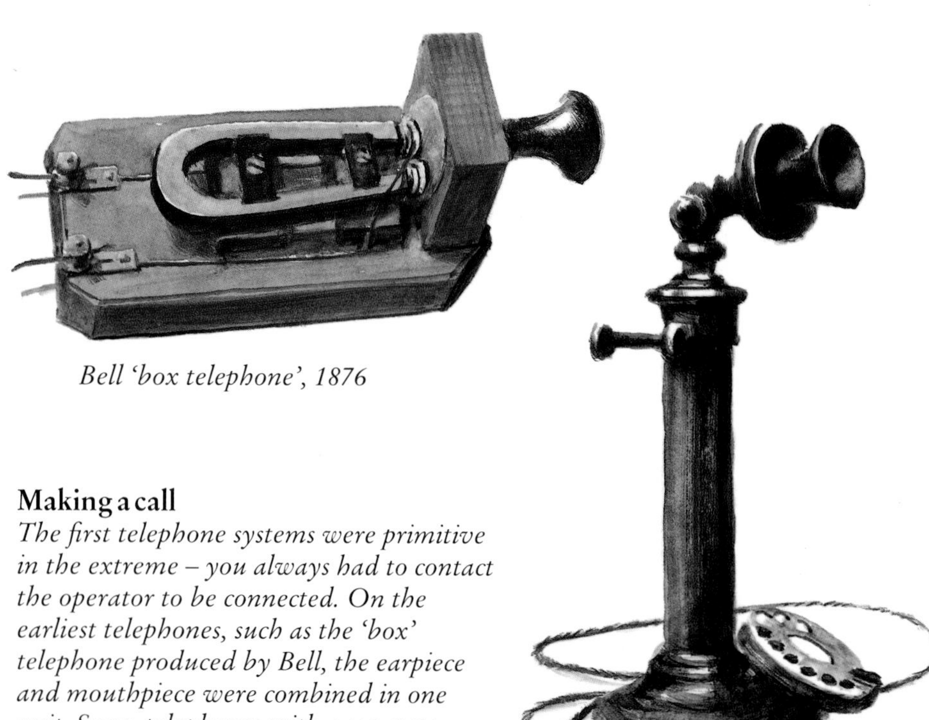

Bell 'box telephone', 1876

Candlestick telephone, 1920s

Making a call

The first telephone systems were primitive in the extreme – you always had to contact the operator to be connected. On the earliest telephones, such as the 'box' telephone produced by Bell, the earpiece and mouthpiece were combined in one unit. Soon, telephones with a separate earpiece and mouthpiece were introduced, and a bell was fitted to indicate that the operator had made your connection, or that there was an incoming call. Only later did telephones with dials, such as the famous 'candlestick' instruments of the 1920s and 1930s, make their appearance.

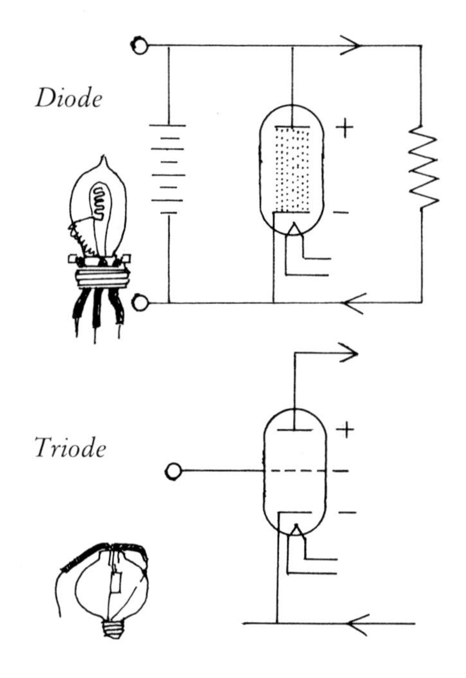

A diode converts an alternating electrical current into a direct one.

A triode can amplify messages and signals. When these amplified signals are combined with radio waves called carrier waves, they can be transmitted over considerable distances.

Wireless communication

By this time the groundwork for electrical communications without wires was being laid. The Scottish physicist James Clerk Maxwell proposed the concept of electromagnetic radiation in 1864. His work was followed by that of Heinrich Hertz, the German physicist, who in 1888 first produced and detected radio waves.

At first, few scientists realized the potential of radio waves for communication. But important steps began to be made. In the early 1890s Frenchman Edouard Branly developed a device called a coherer, which provided a way of detecting the effect of radio waves.

But the crucial work was done by an scientist unknown at the time, Guglielmo Marconi, who worked at first in isolation, in the attic of his parents' house outside Bologna. He began indoors, using a coherer to detect radio waves from a simple transmitter. He soon discovered that he could increase the distance of transmission if he used an aerial and an earth on both transmitter and receiver.

Working with his brother, he made transmissions in morse code further and further away from the house, until he eventually succeeded even when there was a hill between transmitter and receiver. Confident that he had an invention of immense usefulness, he approached the Italian government, who were not interested. Marconi instead turned to Britain (he was half-British and had contacts in the Admiralty), where he received more support.

Marconi gradually improved his equipment and increased the distances of transmission, until in 1899 he was able to transmit a message some thirty-one miles (fifty kilometres) from Wimereux near Boulogne, to Dover on the other side of the English Channel. On 12 December 1901 he triumphantly claimed to have sent a morse signal across the Atlantic. Although there is no way of proving this claim, it is scientifically possible. In any case, it is true that soon after this date commercial radio telegraph services were beginning, and that by 1908 Marconi had set up a commercial radio telegraph service across the Atlantic.

Marconi's equipment was highly effective, and quickly proved itself, particularly in the relaying of messages about maritime disasters. But there was much scope for the equipment to be improved and made cheaper. Perhaps the most important innovations were the adoption of thermionic valves (tubes) in radio equipment. The diode, introduced by John Ambrose Fleming in 1904, was a valve with two electrodes, which proved useful as a more sensitive detector of radio waves. The triode (which had a third electrode) was patented by Lee De Forest in 1906, and was adopted as a means of amplifying weak currents. These simple devices were cheap and easy to make, and did much to make radio more accessible.

The increasing accessibility of radio worked in two directions - as a medium of individual communication and as a means of mass communication, the reception limited only by the availability of receiving equipment and frequencies. And the distances that the messages could travel, as Marconi had shown, were vast. The telephone would be limited for many years to communication across land – the first transatlantic telephone service between London and New York did not start until 1937. Meanwhile, radio, and its natural descendant, television, had already conquered these distances. In our own times, communications satellites have made possible the linking of yet more distant points on the Earth, and even with craft in space.

Guglielmo Marconi with his equipment for making radio transmissions. This technology was improved by inventions such as the diode and triode (far right).

SOUND RECORDING

Although it has many practical applications, sound recording is perhaps most widely appreciated for bringing the joy of music to millions of people

It was in the middle of the nineteenth century that the first serious experiments were made with storing sound on a solid medium, and then reproducing it. Among the first to try this was the French engineer Léon Scott, who published a description of a 'phonautograph' in 1856. His idea was to store the sound on a solid substance covered with lampblack. The sound was represented by lateral undulations in a line.

Scott's concept was taken up by a French engineer called Cros, in 1877, who designed a system using a disc with a spiral groove. Steel copies could be made from the original disc, and these could be played back with a metal stylus connected to a diaphragm.

During the same year, American inventor Thomas Alva Edison turned his attention to sound recording. He had been working with Morse code messages, and was recording these on waxed paper tapes; he noticed that when he pulled the tape rapidly past a spring, the spring made sounds of different pitches.

Edison connected this phenomenon with the recently invented telephone, which used a diaphragm that vibrated in response to human speech. Edison realized that he could connect a mouthpiece diaphragm to a stylus to cut a groove in a solid material. A second stylus, attached to an earpiece diaphragm, could then 'decode' the sound from the groove, and play it back to a listener. In July 1877 Edison tried out the idea, shouting 'Halloo' into his experimental recording machine. The device managed to shout a barely recognizable 'Halloo' back at him.

Cylinders and discs

Edison continued to work on his recording system, and by the end of the year a machine he called a phonograph had been made in his workshop. This used a stylus to cut a groove in a rotating cylinder covered with tin foil. The sound waves were recorded not by lateral undulations, but by the varying depth of the impression of the stylus, which cut a series of 'hills and valleys' in the tin foil. When the recording was made, a second stylus, connected to a listening tube, was moved across the cylinder to play back the recording. Edison's first recording was the nursery rhyme 'Mary had a little lamb'.

Other inventors improved on Edison's phonograph. Americans Chichester A. Bell and Charles Sumner used wax cylinders, which gave better sound quality. German Emile Berliner designed a system using a spiral groove on a flat disc. In Berliner's system, which appeared in 1900, the disc was made of zinc, and the surface was covered with a fatty material. A spiral groove was traced, and acid used to etch away the spiral in the metal. This was then treated as a master, from which a negative was made. Numerous records could then be pressed from this negative copy and played back.

Thomas Alva Edison

In the following years, Berliner's system caught on, and was improved in various ways – with better thermoplastic materials for the discs, with the use of electric motors to turn them at a constant speed, and with the design of horns for use in both recording and playback. A great step forward came in the second decade of the century, when it was discovered how to make an electrical amplifier with valves (tubes). Electrical equipment made recording sessions much more straightforward – no longer did musicians have to huddle uncomfortably around a large acoustic horn. It also led to much-improved sound quality.

The same principles were used in recording until comparatively recently. Vinyl discs replaced shellac, 78 revolutions per minute (rpm) discs by 33 rpm ones, and stereophonic sound offered a leap forward in reproduction quality. But the concept of a spiral groove carrying a continuously varying analogue representation of the original sound remained the most widely used method of sound reproduction until the late 1980s.

The most significant breakthrough of recent years has been the storing of sound as a digital code that can be translated back into analogue signals via a small computerized player. The compact disc and other digital sound-carriers like the Digital Compact Cassette combine high-quality reproduction with ease of use. They continue to make more and more music, at vastly improved fidelity, increasingly widely available.

On an early phonograph a horn was used both to gather the sound when recording and to amplify it when it was played back. This made it relatively easy for a single sound source, such as the human voice, to be recorded, less easy for a group of instrumentalists, who had to crowd around the horn. At the top of the picture is one of Edison's early sketches for his phonograph, showing a needle making a trace in a cyclinder covered with tin foil.

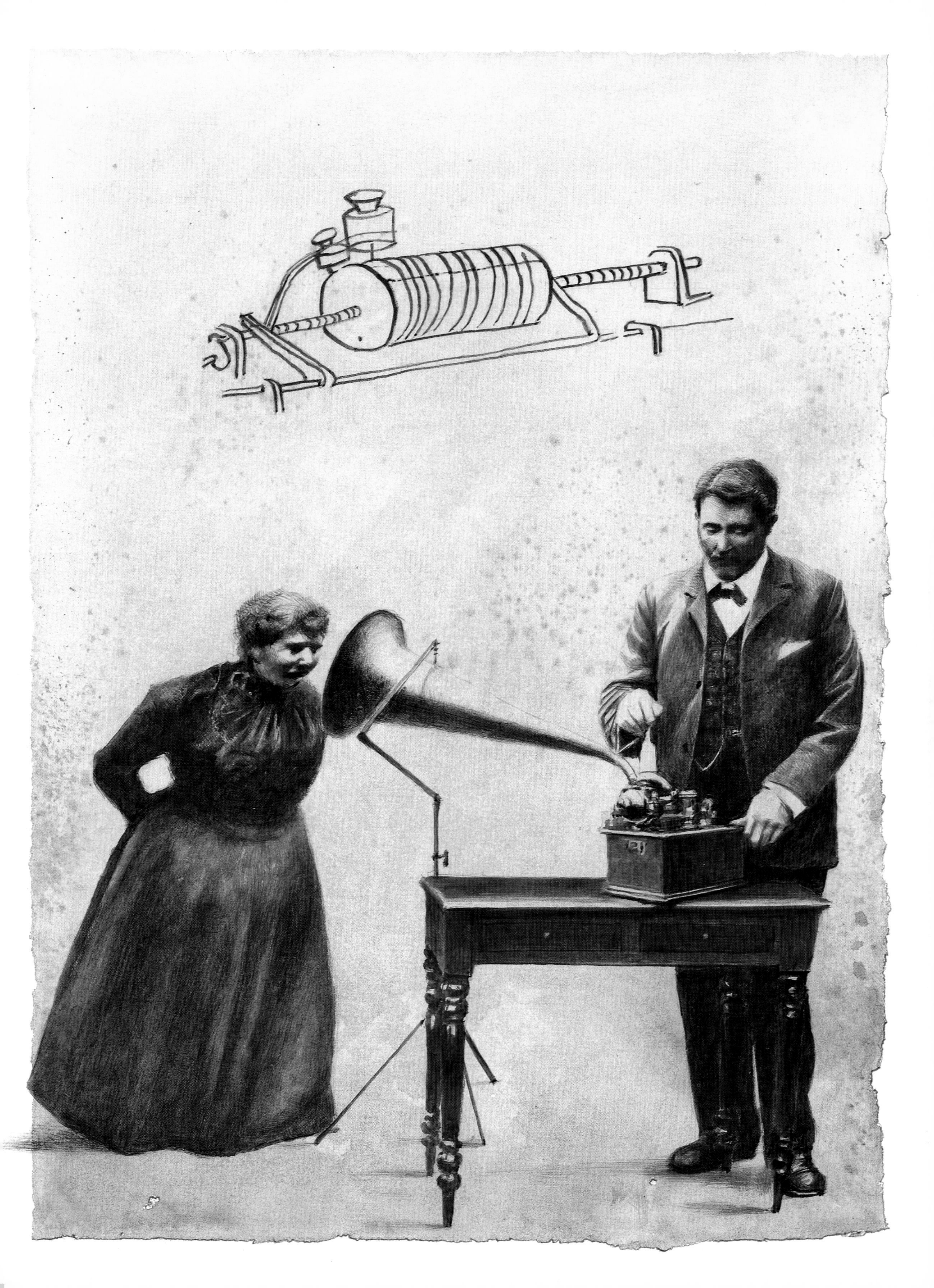

MOVING IMAGES

Early television pictures were of such poor quality that performers had to wear clown-like make-up to give their features adequate definition

Like so many inventions, the cinema owes its existence to the collaboration of a number of different ideas and technologies. Some of these involved the analysis and understanding of motion itself, and how it is possible to break an event up into a series of still images which, if viewed in a certain way, will create the illusion of movement. Some involved the development of photography – necessary to capture the images on film. Others centred on the optical developments needed to reproduce the images. And yet others concerned the development of projection equipment to make the images both large and bright enough to be seen by many people at a time.

Persistence of vision

It is clear from this outline alone that no one person could have invented the cinema. Motion pictures evolved over a period of time, in workshops and laboratories scattered far and wide in Europe and America. One of the first scientists to contribute to the history of the cinema was Frenchman Joseph Antoine Ferdinand Plateau, who made public his concept of persistence of vision in 1829. What he had discovered was that if you break down a movement into a series of 'stills', and then play these back in rapid succession over the same time span as the original action, the viewer's eye will be tricked into seeing the result as a moving image. The human eye will retain an image for only about one-thirtieth of one second, so the individual stills merge together into a moving sequence.

Plateau even made a device to demonstrate his discovery. This was called the Phenakistoscope, and consisted of a spinning, slotted disc bearing numerous successive stills of a moving object. As the disc spun, the image seemed to move – and because the last image on the disc merged with the first, the motion carried on as long as the disc was spinning.

The Phenakistoscope spawned numerous improvements and imitations. All of these were based on the phenomenon of persistence of vision, and they all worked in a similar way, with an 'endless' sequence of step-by-step images. By the mid 1860s, similar devices, like the Zoëtrope (invented by Englishman W. G. Horner in 1834) were being produced commercially. These allowed the user to interchange different strips of images so that different movement sequences could be viewed. What started as a scientific demonstration was fast becoming a parlour amusement for the middle classes, and people were getting their first glimpse of the fascination of the moving image.

Eadweard Muybridge

The photographers

The images reproduced by the Zoëtrope and similar devices started out as hand-drawn or printed approximations of moving subjects. It was not until the science of photography had developed, with the work of men like French photographer Louis Jacques Daguerre and Englishman William Henry Fox Talbot, working in the 1830s and 1840s, that accurate 'phase pictures' of moving things were really possible. No one knew exactly how many types of movement really happened until they were analysed photographically. The pioneers who were most influential in this field were the American Eadweard Muybridge, and the Frenchman Etienne Jules Marey.

Muybridge was an eccentric with a vision. Born in Britain as Edward James Muggeridge, he emigrated to the USA and changed the spelling of his name to a version that he felt emphasized his Anglo-Saxon ancestry. He was something of a fish out of water, but he was also a great still photographer who produced ground-breaking studies of human and animal movement. He used complex and unwieldy equipment to produce vast numbers of sequential images of moving people and animals. These pictures showed more than any others that photography held the key to the future of the moving image. He also projected his images, using a machine of his own design, called the Zoopraxiscope, on a lecture tour of Europe in 1881.

Marey saw Muybridge's images and realized how close they were to his own interest in analysing movement. Soon he had developed a 'photographic rifle', a portable device that took twelve photographs per second of subjects like birds in flight. He also moved from glass photographic plates to rolls of film, making the projection of sequences much more straightforward.

Making the connections

Muybridge and Marey worked during the 1870s and 1880s on their respective studies of motion. Meanwhile, George East-

The magic lantern was the precursor of movie projectors. Above is a spinning phenakistoscope disc bearing some of Muybridge's frames of a running horse, one of the earliest moving images to be made.

The horse 'Mahomed'
travelling at about

Some of the key elements of the cinema: a diagram of Lumière's projector shows how a flask of water acted as a condenser lens and heat filter; Thomas Edison is shown with a phonograph – he was interested in sound from the beginning; Chaplin reminds us that cinema would not have survived at all without the real creators – the actors and producers.

man was pushing forward the development and accessibility of still photography, bringing out the easy-to-use Kodak camera, and introducing roll-film in the 1880s. It was also during this decade that another important pioneer became involved in the prehistory of the movies – American inventor Thomas Alva Edison.

Edison is one of those inventors who have become the subject of legends. With well over one thousand patents originating in his laboratory at Menlo Park, New Jersey, he has often been portrayed as a man of almost superhuman scientific powers – indeed, he used to be known as the 'Wizard of Menlo Park'. But Edison had only a limited grasp of science. He relied on a flair for ideas, a large back-up team, and a talent for exploiting his inventions commercially.

Edison had been working on sound recording. He had already had success with his Phonograph, the forerunner of the gramophone, which he conceived primarily as a dictating machine for use in business. But he wanted to increase the scope of his product by adding moving images to the recorded sounds. With this in mind, in early 1888 he talked to Muybridge about combining the Phonograph and the Zoopraxiscope. By the end of the same year Edison's team had come up with their own device for viewing moving images. And in true Edison fashion they moved quickly to file a patent. Their invention, the Kinetoscope, made a primitive type of motion picture available to the public for the first time.

The coming of the peepshow

The Kinetoscope was a far cry from the modern movies. Edison's team used a camera somewhat like Marey's, but with an important difference. The French pioneer had had difficulty moving film steadily but quickly through the camera. The Americans had their film perforated along the edge with holes that engaged with a sprocket mechanism in the camera and the playback machine. It is a solution that has remained in use to this day.

To view the film the celluloid was placed in the Kinetoscope, a wooden box containing a battery-powered motor, a lamp, a magnifying lens, and an eyepiece.

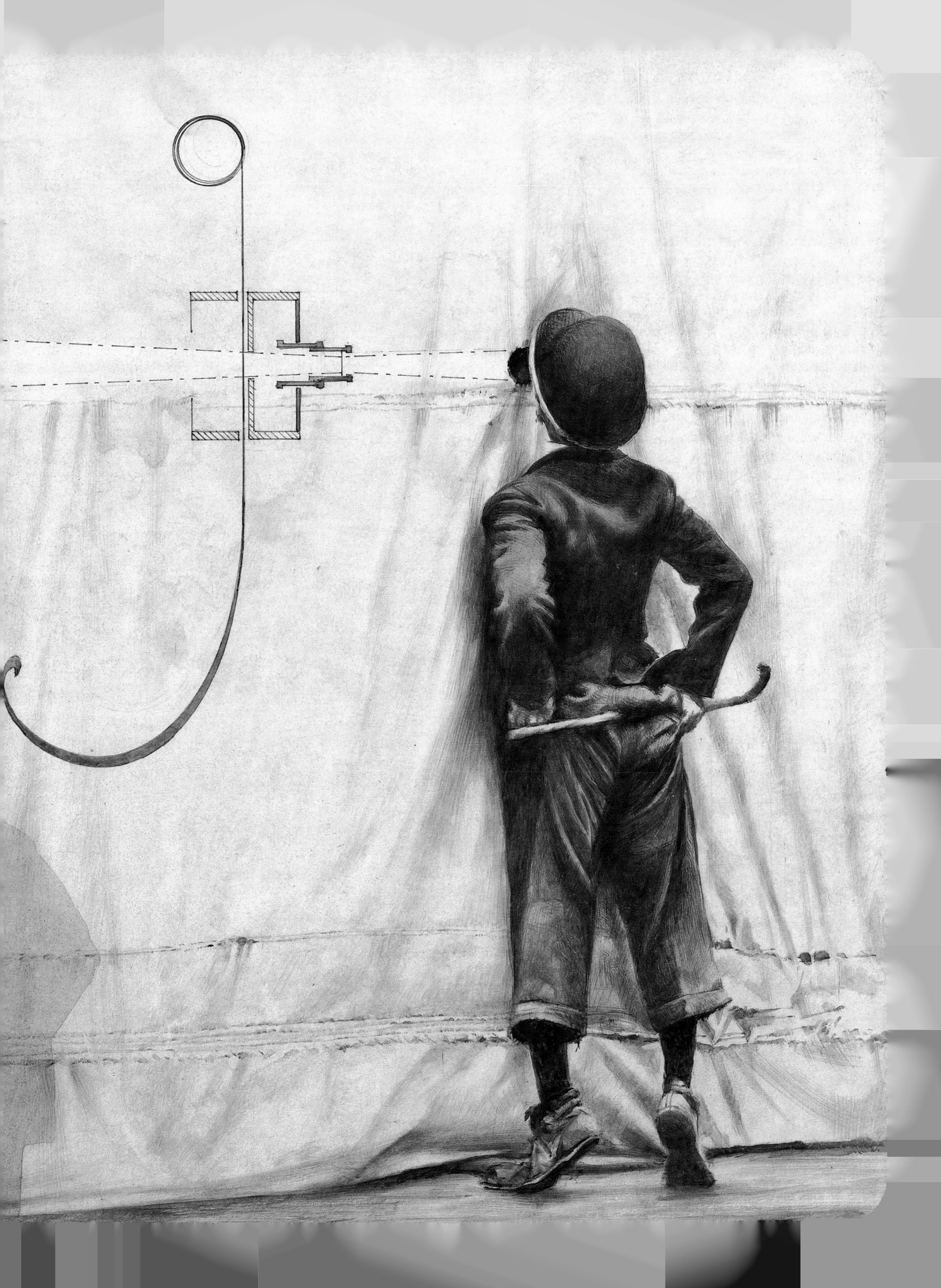

As the film was wound past the lamp at a rate of about forty frames per second, the viewer peered through the eyepiece at the top of the box and saw a short, flickering movie.

Soon Edison had created the world's first movie studio in an attempt to create better conditions in which to produce these films. It was a curious open-topped structure on wheels, which could be turned to follow the sun. It was constructed of timber and tarpaper, which made it lightweight, so that turning was easier still. There was no elaborate equipment inside – just Edison's camera and a basic stage.

Even with the help of the studio, the images on the early films were poor, the films themselves were severely limited in length (some were only a few seconds in duration), and only one person could watch at a time. But people had seen nothing like the Kinetoscope before. The 1890s saw a vogue for peepshow parlours filled with Edison's viewing apparatus. This made money for Edison, but it also inspired other inventors to do better. The obvious need (obvious because magic lanterns had been entertaining people with still images for years) was to transform the viewing machines into projectors, producing large, impressive images on a screen that could be seen by many people at once.

1829 **France** Concept of persistence of vision described

1834 **England** Appearance of Horner's zöetrope

1881 **America and England** Appearance of Muybridge's zoopraxiscope

1882 **France** Marey develops his 'photographic rifle'

1888 **USA** Thomas Edison's kinetoscope patented

1895 **France** The Lumières' cinématographe shows films in public

Towards the silver screen

Edison himself started to develop ideas for projection systems, but apparently lost interest. So the focus switched back to Europe, this time to Lyons, in France, where the brothers Louis and Auguste Lumière ran a factory producing photographic equipment. The Lumières were well placed to develop an interest in motion pictures – they had all the hardware to hand and were used to thinking in terms of light and lenses, photographs and film. So, during 1894 and the early part of 1895, they worked away on developments of the Kinetoscope.

The final key inspiration came about in a single night in February 1895. Louis Lumière, who was prone to migraines, awoke from his interrupted dreams to the inspiration that all inventors crave but few actually achieve. He had conceived a machine that combined the functions of movie camera and projector, a machine that would change the way millions of people saw themselves and the world: the Cinématographe.

In little more than a month the brothers were ready to show their first film. Subsequent showings were arranged throughout the year as the learned societies, specialists and scientists of France were invited to inspect and admire the new invention. But the significant date in the history of the Cinématographe was 28 December 1895, when films were projected before a paying audience for the first time, in the first cinema open to the general public – the converted basement of a Paris café.

The response to the movies

There was little advertising for the early film shows, and few, apart from those who had been to one of the private showings, can have known exactly what to expect. But the Lumières' movies quickly caught on. By the time the brothers put together a proper catalogue of films in 1897, there were already 358 short movies available; by 1901 there would be 1,299.

Many of these early efforts had no more pretensions than the home movies of today. The Lumière's first film had shown workers leaving their factory in

Walt Disney with Pathé equipment

Lyons; later popular successes included footage of a baby eating breakfast, and of someone mistakenly spraying himself with water from his own hosepipe. But they moved. And their very closeness to everyday reality was what was fascinating to the first audiences – they were seeing themselves afresh and were gripped by what they saw.

Other pioneers and entrepreneurs were quick to take up the idea. Charles Pathé, for example, went into business to manufacture film-making equipment in 1896, but soon set up another company to make films. His company went from strength to strength as classic films seemed to pour from his cameras. Many other companies followed in Pathé's wake: the film industry had been born.

And the enthusiasm for movies was not confined to Paris, or even to France. Soon it spread to America, where projected films quickly eclipsed the old Kinetoscopes which had whetted the public's appetite for something better, and where a vast potential audience quickly stimulated the setting up of a huge film industry.

It is impossible in an account such as this to do justice to the many pioneers and inventors who helped movies into being – there were so many individuals who made important contributions. There was the mysterious French inventor Louis Le Prince, the first to design separate apparatus for taking and projecting moving pictures, and who disappeared without trace just as his work was about to bear fruit in

1890. There were also the German brothers Max and Emil Skladanowsky, who had managed to project films by the time the Lumières were showing theirs, but whose projection system did not catch on. And there was the Englishman William Friese-Greene, who was a pioneer of camera design. Any film-maker will insist that no movie can be made without the collaboration of a group of people whose diverse skills are needed to shape and produce the whole. In fact, this combination of inputs was present in the making of movies from their earliest infancy.

'Seeing by electricity', the beginnings of television

A new context for the moving image was suggested when the telephone appeared in 1876. If it was possible to send sounds along a wire, could pictures, even moving pictures, also be transmitted? An answer to this question was sought by several different scientists. A key problem seemed to be how to turn the picture into some sort of continuous sequence of information – to carry out, in other words, the process we now calling 'scanning'.

Some of the most important pioneers looked for an electronic solution to the scanning problem, one based on the cathode ray tube, a vacuum tube that converts electrical signals into visible form by projecting an electron beam on to a screen. Among the inventors who worked on this method of scanning were Boris Rosing, who was working in Russia during the early years of the century, and the British engineer A. A. Campbell Swinton.

Campbell Swinton had worked out a design for an electronic television system using cathode ray tubes in both camera and receiver by 1911, but he did not develop the idea. About a decade later the concept was taken up by a Russian emigré working in America, Vladimir Zworykin. He had worked with Rosing some years earlier, and had got as far as constructing a model of his television system by 1923.

The work of Campbell Swinton and Zworykin was of vital importance but, meanwhile, a British inventor, John Logie Baird, was taking television in a completely different direction. Baird developed a mechanical scanning system based on a rapidly spinning disc pierced by a spiral of holes. The idea of the disc had come from an earlier pioneer, the German Paul Nipkow. The spinning disc system scanned more slowly, and would produce lower-resolution images than the electronic systems, but Baird persevered with it and, for a while, it caught on. In October 1925 Baird showed his first, flickering, low-resolution television picture.

Baird's scanning apparatus
John Logie Baird used the mechanical system of image scanning, in which the picture was broken up by a spinning disc with a spiral of holes cut into it. The inventor donated a piece of equipment designed to do this, like the one shown below, to the Science Museum in London, but there is some doubt about whether it could ever have worked. The original subject on Baird's disc was a moving puppet head.

During the next four years Baird refined his television system, eventually producing images made up of thirty vertical scanning lines and 12.5 pictures per second. When one considers that modern television pictures are scanned at 625 lines, the low definition of Baird's pictures can be imagined.

In addition to his dogged determination to carry on with mechanical scanning, Baird had a talent for publicity. He waged a campaign of demonstrations, and persuaded the BBC to start transmissions from London using his system. The first experimental broadcasts began in 1929.

Baird's pictures were poor in quality, but they were all anyone had see, so they generated much excitement. By 1934 other systems were suitably advanced for the BBC to lay much more exacting requirements for the new public service it proposed to launch in 1936. A standard of 240 lines was suggested by the broadcasting authority, but the EMI company, which had been developing electronic scanning, pledged to do better than this, and develop a 405-line system. For a while, a joint Marconi-EMI 405-line electronic system shared the airwaves with Baird's improved 240-line service.

John Logie Baird

The difference in quality was obvious. There were also many problems for the broadcasters with Baird's system. The EMI television camera could be moved; Baird's was fixed, and anyone whose picture was being broadcast had to sit on a chair with a screw adjustment (and if necessary on a heap of telephone directories) to get their head to the right height. In addition, the flickering light from Baird's mechanical scanner all but blinded the subject during transmission. Another drawback was the peculiar clown-like make-up that everyone on camera had to wear. Finally, continuous breakdowns signalled the end for Baird's system.

And so, mechanical scanning was dropped in Britain within three months of the start-up of transmissions. Other countries where television began early had to face the same dilemma: Germany began broadcasts based on mechanical systems in 1935; in Russia electronic systems were used from 1938. Television in the USA was pioneered by the RCA Corporation, who were fortunate enough to employ Zworykin. His cameras and receivers were using a 441-line system as early as 1937.

Television, then, returned to the electronic roots that had been nurtured by Campbell Swinton and Zworykin. Baird's system disappeared, but its inventor, who achieved those first dim pictures in 1925, has ever since been credited with the invention of television.

COMPUTERS

After labour-saving devices to save physical energy came machines to save mental energy, ostensibly freeing people to be more creative

For centuries, people have looked for ways of relieving the tedium of mathematical calculations. The Ancient Chinese and Japanese used the abacus as an aid to calculation, but it was not until much later that people began to invent machines that did the actual work of calculation for them. Probably the first to do so was the seventeenth-century French philosopher Blaise Pascal, who created a mechanical adding machine, apparently to help his father in his work as a tax official. In the same century, the German mathematician Gottfried Leibnitz improved on Pascal's idea, making a machine that would add, subtract, multiply and divide.

Such machines were not manufactured in large quantities, but their very existence indicated a pressing need. What people needed most was something to help them in the more difficult types of calculations, such as long multiplication and long division. One such aid was provided by Scotsman John Napier, again, in the seventeenth century. His invention, known as 'Napier's bones', was a sort of movable multiplication table engraved on a series of square-section metal rods. When these rods were aligned in the correct way, long multiplication could be achieved by a series of simple additions. Napier's more enduring contribution was his discovery of logarithms, which could be used both to transform multiplication into addition, and division into subtraction.

Babbage's 'engines'

Such devices as those of Napier and Leibnitz were not the true ancestors of modern computers. However, their use did demonstrate a need for help with calculation. One inventor who tried to fulfil the need, with a kind of ancestral computer, was the Englishman Charles Babbage. It was in the 1820s that Babbage embarked on his first machine, the 'Difference Engine', which in the following decade led to the still more ambitious 'Analytical Engine', a universal, automatic calculating machine.

Charles Babbage

Babbage built finished versions of neither of these machines. We owe our knowledge of them to fragments he left behind, which are preserved in the Science Museum in London, and to an Italian account translated and annotated by Babbage's colleague, Ada Countess of Lovelace, mathematician, and daughter of the poet Byron.

Although Babbage was unsuccessful in building his machines, the designs incorporated many important features of later computers – albeit in a mechanical rather than an electronic form. The Analytical Engine, for example, had two main sections: the 'mill', where the actual calculation was done; and the 'store', where the data and the results were kept. Data was input by means of punched cards, which Babbage adapted from those used to control automatic weaving looms. The inventor also distinguished between two types of cards, those containing data to be worked on, and those containing sets of instructions. In these ideas it is not difficult to see the ancestors of many later computing concepts – for 'mill and store' read 'central processing unit and memory'; for 'punched cards' read 'input device'; for 'instruction cards' read 'programs'; and so on.

Equally important were some of the theories behind the instructions that were to be used in Babbage's machines. Lady Lovelace, for example, perceived how a recurring cycle of instructions would often be needed to instruct a computer. She saw that such recurrences could be written into the program in such a way that the machine would 'jump' to the repeated set of instructions if required. This meant that it was not necessary to punch many cards with the same instructions, thus saving much time; a concept still used in computer programming.

Babbage was ahead of his time. People used to think that his machines could not have been built with the technology of the time. This is probably untrue; his machines were not built because Babbage was always devoting his energies to thinking up the next machine before the current one was produced. As a result, his ideas were not widely publicized, and the implications of his machines were not taken up until much later. Meanwhile, simpler mechanical adding machines for basic calculations were created, and these took some of the strain out of calculation.

Early calculators ranged from the ancient abacus (top), *through Napier's bones* (centre) *to Blaise Pascal's mechanical calculator* (bottom).

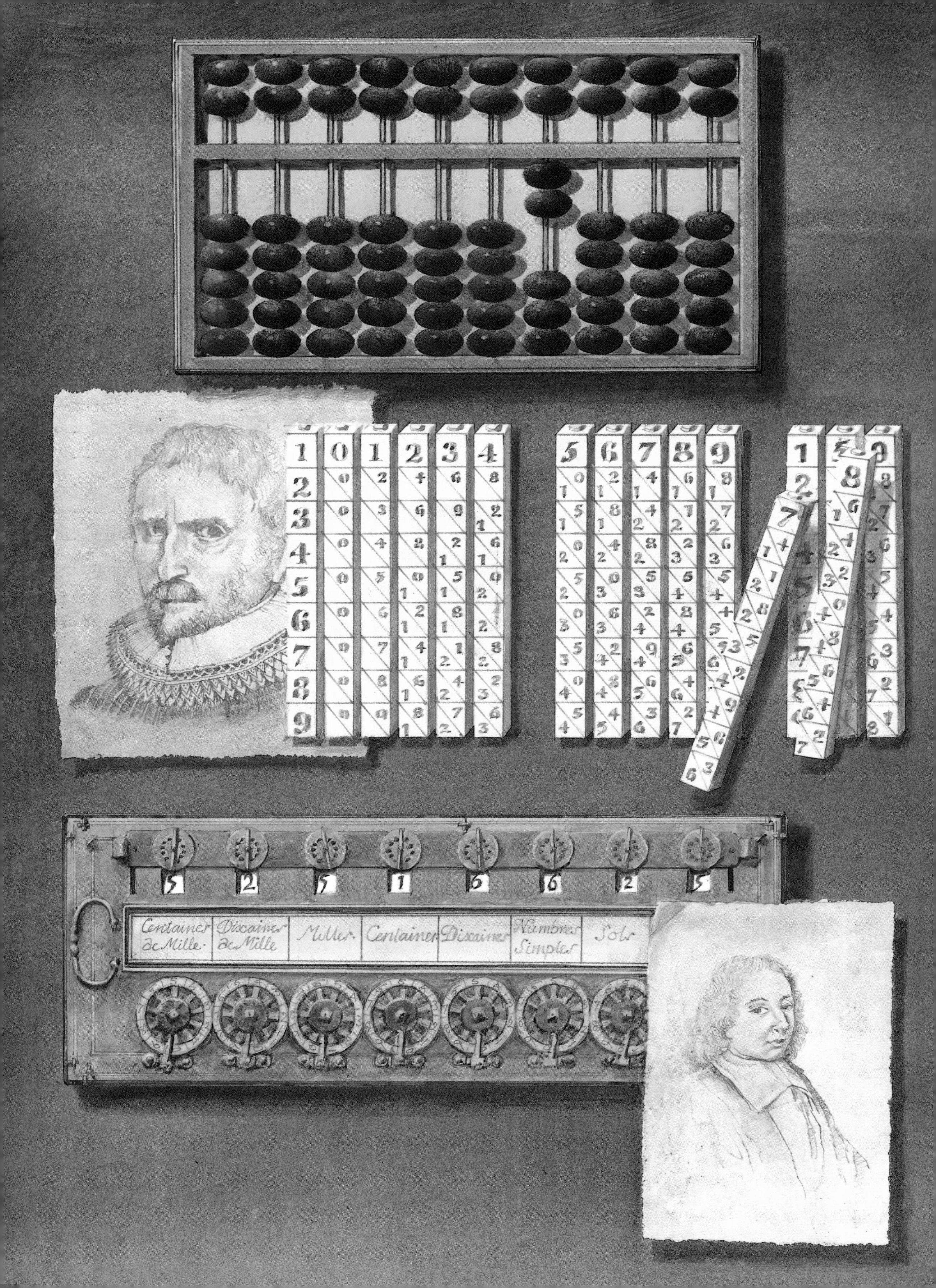

5 2 5 1 6 6 2 5
Centaines de Mille
Dixaines de Mille
Milles
Centaines
Dixaines
Numbres Simples
Sols

Modern computer pioneers

Some of Babbage's concepts were taken up towards the end of the nineteenth century. American statistician Herman Hollerith, for example, used punched cards and paper strips in a machine designed to analyse the returns from the US census of 1880. Subsequent censuses, with the vast amount of data involved, stimulated further attempts at automating the calculations involved. But the development of the modern computer did not really gain momentum until just before the Second World War. Then, as so often, the pressures of conflict helped to accelerate technological change.

Alan Turing

In the mid 1930s, two groups of people on opposite sides of the Atlantic began the work that was to lead to the modern computer. A key member of the first group was the British mathematician Alan Turing, who published an influential paper in 1936. In this theoretical work Turing described a universal computer, the sort of problems it might be used to work out, and the possibility that machines might teach themselves through a process of trial and error. Turing's theory was turned to practical effect when he joined the British Intelligence Service and began to work on the construction of a computer that would be used in the cracking of enemy codes during the Second World War. Meanwhile, in the USA, Howard Aiken of Harvard University was beginning to build an enormous electromechanical calculator with the assistance of the International Business Machines Corporation.

These two endeavours resulted in the world's first two true computers, the British Colossus and the American Automatic Sequence Controlled Calculator (ASCC, also called Mark I). As the name of the latter reminds us, both of these were calculating machines, designed basically to perform quite simple mathematical tasks, but to do so far faster than could any human mathematician.

In fact, the calculation times seem painfully slow by the standards of modern computers, but they were soon to speed up. For example, the first ASCC took 0.3 seconds to add two numbers; the Mark II model, which appeared in 1947, took only 200 milliseconds. This meant that calculations that would take humans years, could be done on a computer in a matter of hours.

Colossus and ASCC represented an enormous technological advance, but they were, in some ways, very primitive, relay-operated machines. A relay is an electrical switching device in which one circuit is controlled by a separate circuit; in the 1940s, this involved many mechanical components. Speed could be improved considerably by using electronic switching, based on the thermionic valves that had been so important in the development of radio and broadcasting.

The first electronic computer appeared in the USA in 1946. It was called ENIAC (Electronic Numeric Integrator and Calculator), and was much faster than its mechanical predecessors, achieving an addition of two numbers in a mere 0.2 milliseconds. Indeed, the main limit on ENIAC's speed of use was the rate at which data could be input and output. ENIAC used punched cards to accept and output data, a laboriously slow process by modern standards. Nevertheless, it could still perform, in a single day, calculations that would take one person an entire year to work out using the kind of mechanical calculator then available. The machine remained in service for ten years.

Key computer concepts

Since the earliest days of computing, even since the machines of Charles Babbage, it had been realized that some sort of storage medium was necessary, in which the computer could keep the data on which it was working. It was also recognized that a computer needed instructions in order to tell it what to do with the data. Computer storage facilities have since come to be known as memories, and sets of instructions as programs.

Programing a computer is a laborious business. The machine needs to be told every step in an operation: nothing can be assumed. For example, it is no use telling a computer simply to add two numbers; it also needs to be told where to display the result so that the operator can read it. Computer programs are therefore detailed series of instructions, which take a great deal of time to write. The instruc-

The Fairchild integrated circuit of 1961 was the first such circuit to be manufactured commercially. It is shown here vastly magnified. To the right are resistors used in modern computers, and a short length of the punched paper tape, which people used to communicate with computers in the 1960s .

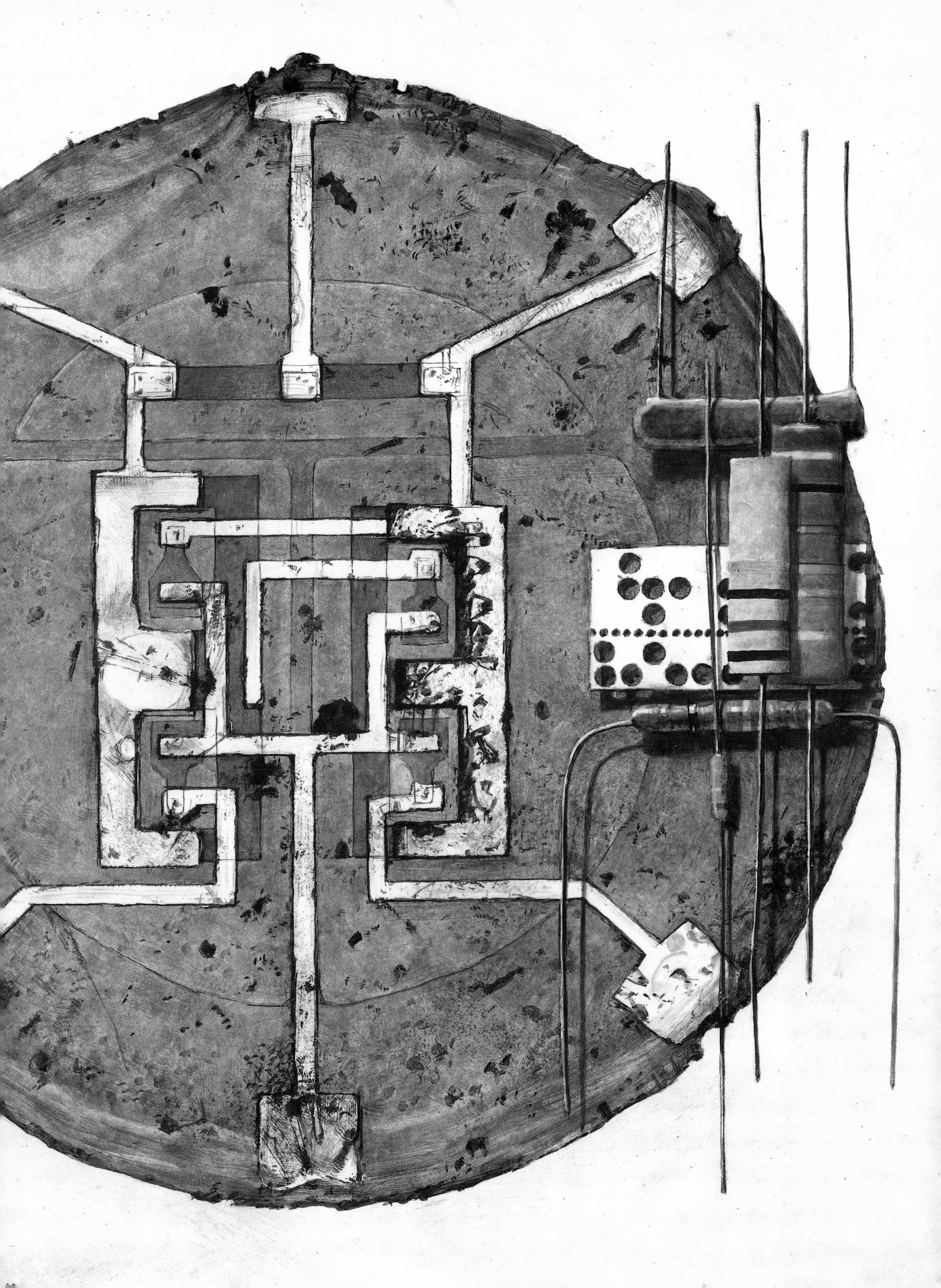

tions also have to be changed as soon as the operation is changed.

It was the mathematician John von Neumann who suggested that the program itself be stored in the computer's memory. He also put forward the idea of a computer that could change its actions according to the results of its calculations.

To achieve this, much more storage space was required than was possessed by the earliest computers. Various types of storage were tried, and one of the most successful was a magnetic drum containing a nickel-plated cylinder fitted with recording heads. These heads could magnetize a track around the nickel cylinder in a pattern corresponding to the numbers to be stored. A magnetic store like this had the advantage over previous systems that it was not erased when the current was turned off. In other words, it introduced the concept of the non-volatile memory, still vital in computers today.

Another of von Neumann's perceptions was the importance of the binary number system for computing. We normally count in the denary (or base ten) system, adding a new digit to the number when we reach ten, another when we reach 100, and so on. Von Neumann saw that the number system using the base two was more appropriate for computers. In the binary system there are only two symbols (0 and 1), so 2 is written as 10, 3 as 11, 4 as 100, 5 as 101, and so on. This is ideal for computers, since the two symbols can be represented easily by the presence or absence of an electrical current. Von Neumann was not the first to see the importance of binary for computers – even ENIAC used binary. But his work, and his influential 1946 series of lectures at the Moore School of Electrical Engineering (part of the University of Pennsylvania) where ENIAC was built, did much to clarify the importance of binary.

Four generations of computer

Binary code quickly became established as the number system in which electronic computers operate. As a result, computers needed numerous internal switching elements with which to turn the current on and off. Computers like ENIAC used some 18,000 electronic valves (tubes) to do this job. By the 1950s, numerous computers containing thousands of valves were still being built – the IBM 701 of 1953, for example, contained 4,000.

Valves were among the most familiar electronic components of their time. They worked, but at a cost. For one thing, the life of a valve was limited to a few thousand hours, so a machine containing numerous valves would be out of operation for a large amount of time. Valves also used up much electricity and generated a lot of heat. This meant that they were costly to run – although if the air-conditioning system was built around the computer, the machine could be used to heat the building in which it was housed. In addition, valves were large and contributed to the cumbersome, room-filling designs typical of early, so-called 'first-generation' computers.

Help was at hand in the form of the transistor. This is a much smaller component of the type known as 'solid-state'. Solid-state components work according to the movement of charges within a crystalline solid. They are lighter, stronger, more robust, and more durable than valves. They were developed in 1948 by William Shockley and other scientists at the Bell Telephone Company, and began to be used in computers in the late 1950s to replace valves; these computers are known as 'second-generation' machines. They were more economical, smaller, and more reliable than their predecessors.

The next stage, the 'third generation', was represented by still more compact computers. They began to appear in 1964, when manufacturers found a way of squeezing a host of components for a specific function on to a single piece (or 'chip') of silicon. The manufacture of the chips was done under controlled conditions and a number of chips could be connected together easily to form the basis of a computer. This made the manufacturing

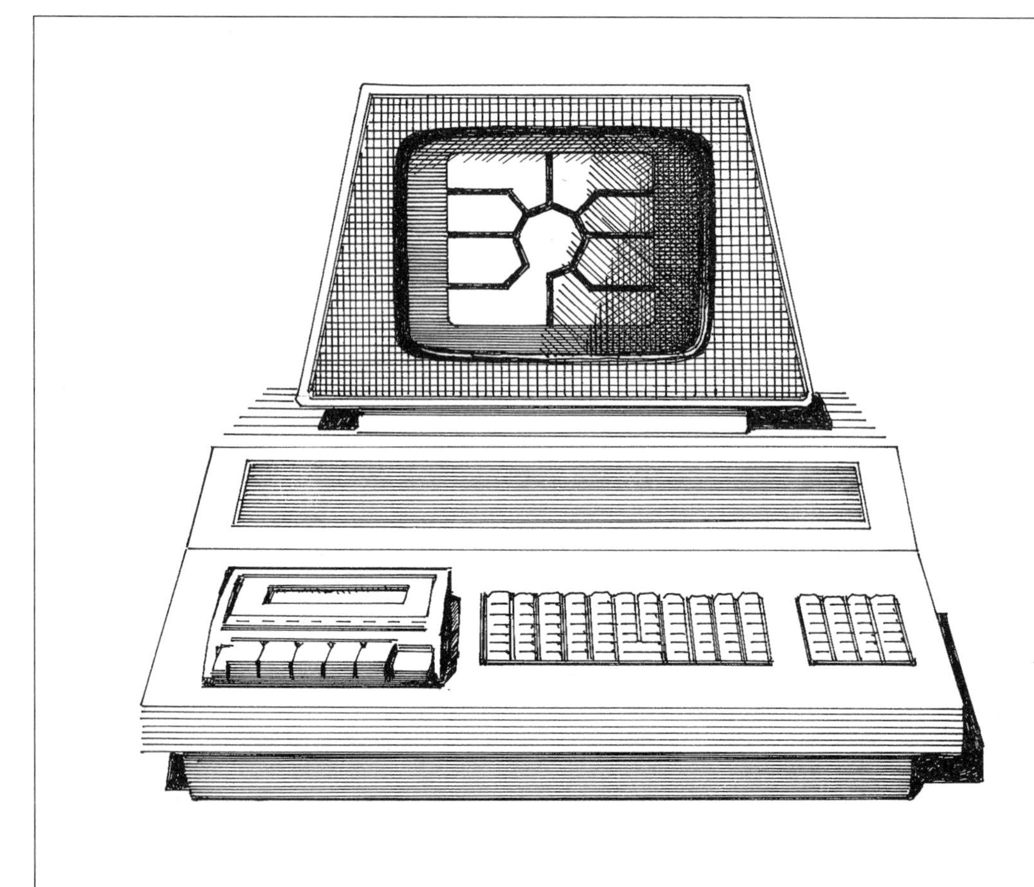

The silicon chip and micro-computer

It is astonishing how small modern computers can be. The computers that sit on desktops all over the world today are more powerful, faster, and far more reliable than their pioneer predecessors, which took up whole rooms and comprised thousands of separate, fragile components. This miniaturization is largely the result of the integrated circuit. In such components, a number of different circuits can be put directly on to a wafer-thin piece of a semi-conducting substance, such as silicon. This is done chemically, in a process combining etching and photography. The result is a series of microscopically-small circuits built up in layers on the silicon, which can house or process large amounts of information. The circuits on the silicon are already connected together, ready to perform the tasks in the computer for which they are intended. They may be designed as memory circuits, or as the chips that make up a computer's central processing unit. Whatever their function, comparatively few of them need to be connected together to make quite a powerful machine.

Stored in code

Although a computer's primary work is calculating, almost any kind of information can be handled and processed in some way by a computer. This is because it is possible to encode any sort of data into a series of binary digits (1s and 0s, or ons and offs), which the computer can then manipulate. Many of today's uses seem a far cry from the calculations for which the first computers were designed. For example, artists and designers can use computers because images can be represented digitally. Architects (and town planners) are increasingly using the power of computers to depict three-dimensional scenes, which can be viewed from any angle. Musicians have also found uses for computers, to write, record and play music that has been digitally coded. Indeed, this is one of the most familiar modern uses of digital technology, since the compact disc stores music in a computer-readable digital code. The information is recorded on the disk as a series of pits that are 'read' optically by a laser.

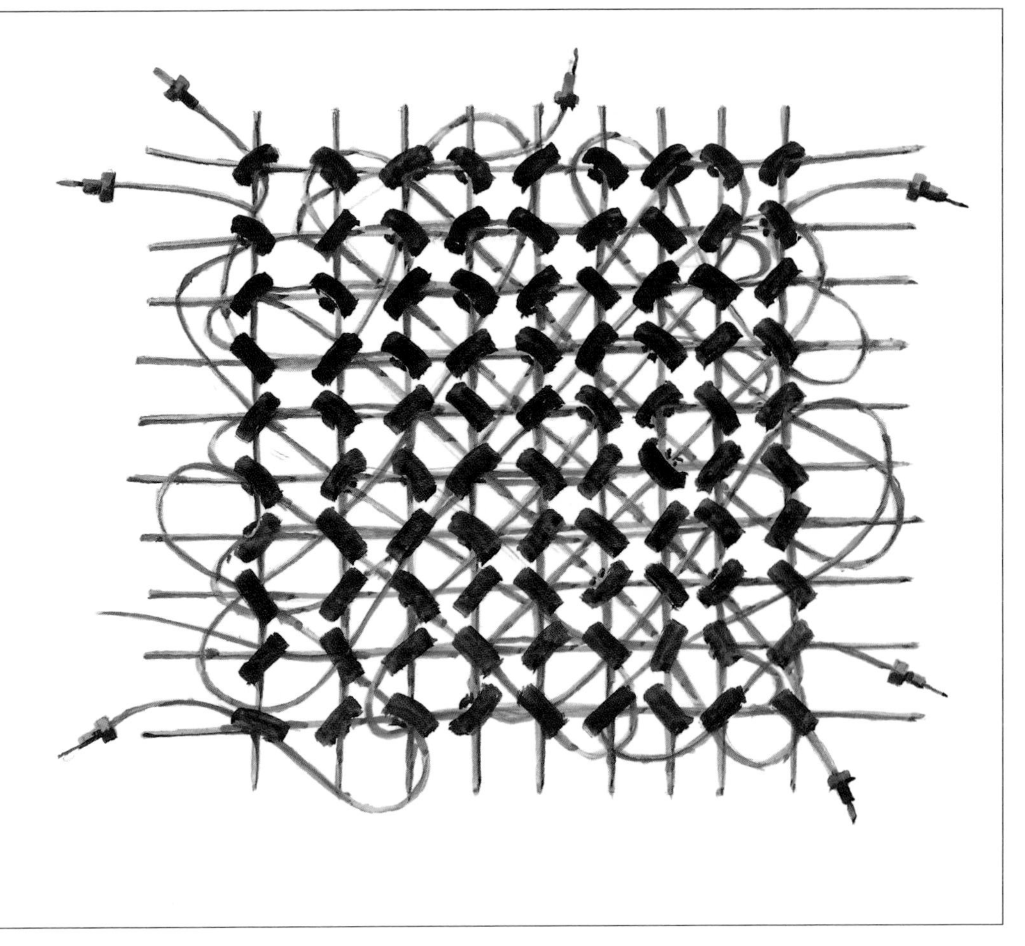

process more reliable, as well as making the finished machines still smaller. The machines were also faster, because the electrical impulses travelled across shorter distances.

In about 1970, the 'fourth generation' of computers started to appear. These machines are built using the technology of 'large-scale integration'. This involves including even more components on the silicon chip, so that virtually the whole central processor of a computer can be contained on just one piece of silicon. With such compact circuitry has come the age of the micro-computer, with machines that will fit easily on to a desk, on the user's lap, or even, most recently of all, in the user's pocket, becoming increasingly common.

A widening influence

What is more, the computer is no longer the preserve of the research laboratory, the government department, or the big industrial corporation. Third and fourth generation computers became increasingly straightforward to produce, as ways were devised of automating the manufacturing process. This automation also helped to reduce errors of production, and computers became cheaper as a result.

And so, during the 1970s and 1980s, the computer became an increasingly familiar and influential tool, affordable by smaller companies, schools, and individuals. Everyday office tasks such as the writing of letters and reports, keeping accounts, and handling 'banks' of information (from simple lists of names and addresses to complicated records) became the undisputed realm of the microcomputer. It has found new roles in medicine and education, and even in fields such as art and design.

Unthought-of changes in the way people work, and in the ways businesses are run have occurred as a result of the use of computers. It is not simply that some tasks, such as producing a neatly typed, clean copy of a letter, are now far easier, and require less labour. Some tasks have been eliminated altogether. Take stock control, for example. By having a computer at the point of sale on which each purchase can be recorded, exact records of stock and sales frequency can be kept automatically. When stocks run low, the computer can print out an order for the required item – or even contact the supplier directly via a telephone line.

As the above example suggests, microchips are used in many places where there is no actual computer in the usual sense of the word. The control systems of a vast range of equipment, from complex industrial robots to familiar appliances in the home, such as dishwashers or telephones, contain microchips. The concept of 'programing' a familiar household device such as a washing machine or a central heating system shows how far into the home environment microchips have already reached. The cheap, highly accurate wrist-watches of today also use microchip technology.

Meanwhile, computers are still being used for the large-scale tasks for which they were originally conceived. In the second half of the twentieth century there have been certain developments, for example in the harnessing of nuclear power and in-space exploration, that have involved calculations on a vast scale. Such developments would have been impossible without computers. Indeed, a whole range of tools, used in fields from industrial design to medical diagnosis, rely for their function on computers.

And if such applications seem remote from our everyday lives, they have their impact in ways that often affect us. The sort of sophisticated monitoring and guidance systems required in spacecraft, for example, have had their influence on the technology of aircraft instrumentation. Many of the minute-by-minute decisions involved in flying a large aircraft can now be taken by a computer.

Talking to computers

This line of development has seen the core (the central processing unit) of the computer getting steadily smaller and more reliable. There has been a parallel development, which has made it easier and easier to get information into and out of a computer. To begin with, entering data into a computer was a tedious, two-stage operation. Information had to be typed on to cards or paper tape, where it appeared in the form of a pattern of holes. The cards or tape were then loaded into a reading unit so that the computer could access the information.

The adoption of keyboards linked directly to the computer made data entry much more straightforward. The keyboard was, and is used in conjunction with a screen. Using this, the data typed in can be checked instantly by the operator to avoid mistakes.

Quite early on in the history of computing, it was realized how easy it would be to communicate with a computer with a familiar and simple-to-operate device such as a teleprinter. In 1939, engineers at the Bell Telephone Laboratories designed a computer that accepted information in teleprinter code. It could therefore use data inserted at any teleprinter connected to it, and could also print out the results to the teleprinter. The first time it was demonstrated, in 1940, the computer in New York was operated from a teleprinter at Dartmouth College, New Hampshire.

There were further advances in the provision of printers for producing hard copy from the computer, and in the use and development of magnetic systems for storing data. The latter were at first based on magnetic tape, later on the magnetic disks that are now the familiar medium for storing programs and data.

All of these innovations have made it easier for people to use computers, to put information in, and to take it out. Still more devices, from pointers that mimic the action of a conventional pen, to touch-sensitive screens, promise to make the machines still more accessible.

But the greatest advances are in the invisible areas of the computer, the programs that make the machine work. These are being designed to make entering and retrieving information so easy that one does not even have to learn how to do it. Instead of typing in complex codes, the user can point to a picture on the screen that represents a particular set of instructions, or even give the computer direct verbal instructions by talking into a microphone. The machine itself translates these commands into the codes it needs in order to carry out the instructions. Such developments have made computers beguilingly easy to use, and have ensured that they will maintain their influence on our lives – an influence perhaps greater than any other idea of the modern era.

Multimedia

One of the most exciting and accessible recent developments using computer technology is what has become known as multimedia. This is the use of the computer's capacity for information storage and retrieval in a special way. As well as words and figures, it is possible to store still pictures of all sorts, sound, and even animations and other kinds of moving pictures as digital codes in a computer's memory. It is therefore possible to create a 'reference book' from a combination of all these kinds of data. Thus, information about, for example, a composer, could be illustrated with extracts from the scores and music examples that one could listen to. An account of the workings of the internal combustion engine could be brought to life with an animated sequence showing the combustion cycle. A sports training manual could 'freeze' a particular movement at any point to show the ideal approach.

The richness of such a method of information-carrying is obvious, but this is only the beginning. The computer's ability to search for whatever data the user specifies allows one to 'read' this kind of electronic book in a new way. Instead of reading in a linear fashion, from beginning to end, or turning continuously back and forth to the index to find references to one's chosen subject, it is possible to let the computer do the searching. So, in the case of a reference work on composers, one could use the search facility to pick up all the examples of symphonies, in chronological order, to provide a history of that musical form. A host of other themes could be picked out in a similar way. It might also be

possible for the user to select the level of information required, with different text available for adults or children, beginners or advanced students. Thus the reader can structure the way in which information is presented, while yet retaining an organized approach, in a way that is well-nigh impossible with even a very well-indexed traditional book.

Another benefit comes from the miniaturization possible with digital technology: a vast amount of information can be crammed into a small space. A common method of storage, for example, is a compact disc (CD) similar to the sort used for recorded music. Such a disc can hold the text of an entire encyclopedia and a large dictionary. Sounds and images take up more digital space, but it is still possible to put a sizeable reference work on to a CD. A computer is still required to gain access to the information on the disc, but, such machines but are steadily shrinking in size.

Multimedia is an ideal tool for a certain type of information retrieval. It is already being used for the interactive cataloguing of large libraries, for example, and in the provision of on-screen 'art galleries' that provide background information about all the paintings. But there are still many functions for which the traditional book is better suited. And screen technology will have to improve considerably before people feel as comfortable with words on a computer as they do with words on a page. Multimedia represents an information revolution, but the book is not dead.

EPILOGUE

The inventions that have been described and illustrated in this book form a series of starting points. They have been developed in countless, often unpredictable directions. None of the mine owners who used primitive steam engines to pump water out of their mines predicted the railways; neither could Stephenson and the other railway pioneers have imagined a Japanese bullet train or a French TGV. And such developments are still going on: the human desire to invent and the range of human needs calling new inventions into being show little sign of diminishing.

The richness of modern inventiveness can be illustrated with one or two well- known examples. Take hand tools. In some ways they have developed very little. A Roman carpenter would recognize our saws, knives, hammers and chisels. But these have also been combined with other inventions, such as improved materials and electric motors, to make power tools, devices that cut, drill or plane with a speed and ease that would amaze our ancestors. Tools have also been taken into new workplaces, and still newer technologies have been enlisted to help them do their jobs. Modern surgeons no longer rely on the metal scalpel for every operation. Where the highest precision is required, such as in eye surgery, sophisticated laser cutting tools allow operations that were unknown before. Another development has seen robotic machine tools used in many of our factories. Such robots can perform repetitive actions automatically under computer control, again with great precision. The production of everything from cars to computers has been transformed as a result.

The other fields covered by this book have seen similar developments. Satellites have dramatically opened up global communications, as well as revolutionizing the science of meteorology. Scanning electron microscopes have provided glimpses of yet smaller and smaller organisms. Radio telescopes allow us to look deep into space. Refinements to the internal combustion engine make it more efficient and less of an atmospheric pollutant. Exciting developments in power generation offer the possibility of energy which will not be a drain on the Earth's resources. Computers continue to be adapted for more and more uses at home and at work. The range and usefulness of new inventions can only inspire admiration for the scientists and technologists who bring them into being.

New technologies, new challenges

But change can cause difficulties that the technologist cannot be expected to solve. For example, we were told some years ago that computer technology would lead to the paperless office; data stored electronically would rule out the need for bulky paper records and thus save resources. But the ease with which hard copy can now be printed out, and the apparently irrepressible human need to see things in print, have meant that computers have heralded an increase in the amount of paper we use.

Another example is the phenomenon of 'homeworking'. With computers, fax machines and other electronic aids it is perfectly possible for many people to work at home, sending data down telephone lines as and when required. Travelling is reduced and again resources are saved. But there is much evidence that people do not want to work like this: they value the society of the workplace and crave person-to-person contact.

We are not ready for every innovation and every possibility, but at least we have the choice. Less fortunate have been people in the developing world who have had 'help' from richer nations in the form of technology that is simply inappropriate. As development workers now realize, technology has to be available and maintainable, and the people who are to use it must be able to understand its potential. Often this means studying the indigenous technology rather than imposing something from outside.

But in spite of the difficulties, new inventions offer exciting possibilities. And there are some emerging innovations that hold great promise for the future.

Power generation

Nuclear power is an important example. There was a period of a decade between the use of the first atomic bomb in 1945 and the first practical nuclear reactor. These devices, as we have seen, work by fission ('splitting the atom'). Unfortunately, power is not generated by nuclear fission without considerable risk to the environment. The process is highly hazardous and those working in nuclear reactors have to observe stringent safety procedures. In addition, there is radioactive waste to dispose of, an environmental problem that we still have not solved.

An alternative, cleaner form of nuclear energy would be far preferable to nuclear fission. The possibility of this seemed to come nearer in 1952 when the fusion bomb (also known as the hydrogen or thermonuclear bomb), was first exploded. Since then, scientists have been trying to create a practical fusion (or thermonuclear) reactor.

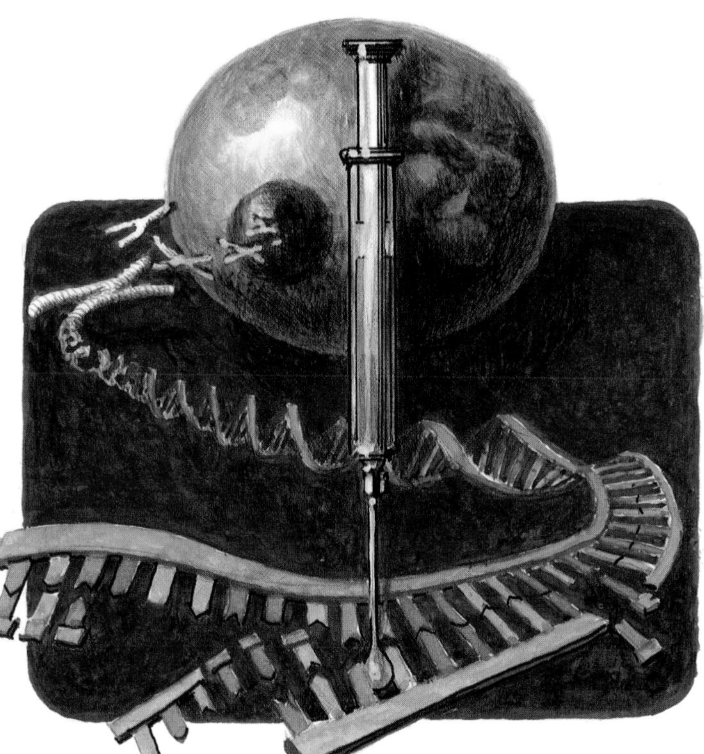

In nuclear fusion two light nuclei (such as hydrogen or deuterium) combine to form a new stable nucleus. This nucleus is lighter than the twin component nuclei together, so energy is released. Nuclear fusion has several potential advantages over fission as an energy source. It requires only inexpensive fuel (which can be made from sea water) and its waste products would be 'cleaner' and easier to deal with than those from conventional fission reactors.

But some 40 years of research have only shown how difficult it is to produce a fusion reactor. For one thing, incredibly high temperatures have to be reached before the atoms can be brought together. This means that any conventional container would melt – scientists may have to use a magnetic field to contain the atoms. And it has proved difficult to build a reactor that will produce more energy than it consumes. One day there will probably be a solution, and cheap, relatively clean energy should be the result, but it may very well not come until the next century.

Meanwhile, some headway is being made with 'renewable' energy sources – tide and wave power; solar energy; and wind power. Individually, these sources have severe limitations. You have to be at a place on the Earth's surface where the elements can be harnessed in these ways. But there are already a number of solar and wind generating facilities (often small ones) reducing the amount of energy required from less clean sources. There is also geothermal energy from heat produced in the interior of the Earth. Another idea envisages algae-fired power stations, and scientists have already found ways of growing algae in vast quantities without expending too much energy, and of storing them as a dry fuel. Such forms of energy offer more environmentally sound solutions to energy requirements than many of those in use at present.

Resources - technology and the planet

The whole issue of the responsible management of the planet's resources is also one that will occupy inventors in the future. Well-known examples are the depletion of the ozone layer above the Earth and the increase in the process of desertification (the creation of new deserts).

Ozone depletion has stimulated people to reduce their use of chemicals known as chloro-fluorocarbons (used widely in aerosols and refrigerators). It has also made scientists come up with alternative chemicals which are gradually (many would say too gradually) being introduced.

Desertification has led to new approaches to land management. Arresting wind erosion, planting drought-resistant species to put organic material back into the soil, irrigation, and making best use of any available water are the processes that need to be undertaken. In addition, new water desalination processes are helping to augment the water supply in many arid regions. One successful process involves reverse osmosis – using a membrane which lets the water through but retains most of the substances dissolved in the liquid.

New technologies are helping in many other areas of the new science of planet management. Typically they involve both the individual and the entire globe. This is not only the case in the negative actions we take to help the environment (such as stopping using aerosols to help the ozone layer). It also involves positive measures. It is to be hoped that more and more people will live in energy-efficient homes that are made with sustainable materials and built to save power and to be recyclable. Recycling of other items that we use, from tin cans to paper, may also become more widespread. And we may be encouraged to use items made of the new biodegradable plastics rather than the older plastics that have posed such long-lasting waste-disposal problems. In short, technology can help solve many environmental problems, provided that people are sufficiently motivated to take up the solutions – quickly.

Genetic engineering

People themselves have been the subjects of some of the most interesting recent research. In 1953 J.D. Watson and F.H. Crick first proposed the structure of DNA, the nucleic acid that forms the main constituent of chromosomes, which carry the information needed for the functioning and organization of all living cells. Since Watson and Crick hit upon the now-famous double-helix structure of DNA, there has been a vast amount of research into the information carried by DNA, how it is arranged and how it affects our lives.

DNA is made up of four chemical bases which can be arranged in an almost infinite variety of ways to make up a sort of 'computer program' that is unique to each individual. The program determines a host of key characteristics, from gender to how tall we are. It is made up of 23 paired elements called chromosomes (we inherit one of each pair from each of our parents). Each chromosome is made up of thousands of individual coded instructions called genes; each gene controls the body's production of a particular protein; each protein controls a particular bodily function.

When a gene is defective in some way, disease can result. As we have pairs of chromosomes, we have duplicates of most of our genes, but illness can be caused if neither gene works properly. Many diseases are inherited and might one day respond to gene therapy. A common genetic disease is cystic fibrosis, which results in the absence of a protein that controls the movement of salt and water across membranes in the body. In cystic fibrosis the lungs produce a mucus that makes breathing difficult and plays host to infections. Cystic fibrosis is just one disease which researchers are hoping to cure by manipulating the patient's genes. Others include Alzheimer's disease, Parkinson's disease and many forms of cancer.

Techniques for introducing new, correctly working genetic material into the body are still in their infancy. One method involves using viruses (which themselves have been genetically disabled so that they cannot multiply) to carry chromosomes bearing therapeutic genes into the body. Another method involves micro-injections of DNA.

All living things contain DNA, which has been descibed as 'the blueprint of life'. If we can read the blueprint, it is therefore theoretically possible to alter it, and thus modify the form of any living organism. The potential here is vast, particularly in areas such as food crops. Scientists anticipate that they will be able to produce fruitful, regularly cropping strains of such plants ideally suited to wherever in the world they need to be grown. The possibilities for reducing world hunger are enormous if those who fund the research are willing to follow it up. But the ethics of 'altering life' in this way need to be discussed. It is a concept with which many lay people feel uncomfortable.

Space travel

After the frenetic activity of the 1960s and the problems with the space shuttle, the public interest in space travel is much less than it was. But there is still much valuable research going on – research which will modify our view of the universe in a number of important ways. Space probes of the 1970s and 1980s have already passed close to most of the planets in our solar system. Such robot missions will certainly continue, and eventually people may one day go to Mars.

In addition, humans will travel into space to build larger space stations than have previously existed. Such space stations will be taken up on reusable craft like the shuttle, and assembled in space. These outposts of Earth will allow a crew to live and work in space for long periods, but will be a far cry from the large space colonies imagined by science-fiction writers. It will be a costly exercise, but the price will be borne by a number of nations, and the resulting hardware will probably remain in the corrosion-free environment of space for decades.

In the near future, then, the emphasis on space research will be on work close to home, but there will eventually be excursions further afield. For one thing, probes designed to examine the outer planets will leave the solar system as they continue on their way. This phenomenon began as early as the 1970s, with the *Pioneer 10* spacecraft.

For space travel to progress in the way that its proponents want, methods of propulsion will need to develop – nuclear power is a possible method. Another potential power source is the fuel cell. This is a device that converts the energy resulting from a chemical reaction directly into an electric current. It provides a clean source of power and, unlike an accumulator, does not have to be recharged.

Single-stage, fully reusable launch vehicles will be needed. What is more, they will have to be able to be controlled remotely from Earth if the full potential of unmanned flight is to be realized. There is a vast amount of work to do even to travel the cosmically short distances within our own solar system. But the human longing to discover and learn about new worlds will mean that the effort and the money will continue to be found for space research.

Developments in computing

The computer is a product of technology that has changed many lives very quickly in recent years. There is every sign that it will continue to do so. Even computer pioneers like Alan Turing looked forward to the time when computers would be able to 'think' in a way similar to humans. In the early 1960s computer programmers tried to apply the concept of artificial intelligence to produce a machine that would translate text from one human language into another. They soon realized that the way we use language is more complex than they first thought – we draw on a huge amount of background information when using language, for example.

Even if it is possible to program some or all of this background information into the computer, the machine takes a long time to process it. This is largely because computers deal with information *serially,* one piece of data at a time. The human brain, on the other hand, is capable of dealing with many pieces of information *in parallel.* For example, while walking from the station to the office, one might be simultaneously planning the day ahead, avoiding other pedestrians, noticing a special offer in a shop window, and putting up one's umbrella as it starts to rain.

We are a long way from machines that can truly think in this way, but this is one of the most promising developments for the future. The applications would be many and diverse. Car manufacturers are positing vehicles that would use their intelligence to help one avoid accidents or losing one's way. Automation would extend into more everyday tasks than is at present possible.

How could a machine mimic such thought processes? One way in which computer scientists are looking at this problem is by connecting simple processing devices together in a network somewhat similar to the network of neurons in the human brain. These so-called 'neural networks' are proving promising in such difficult areas as word recognition, in which a computer is required to recognize spoken words and display them on its screen.

But word recognition is a simple task compared with what the human brain is capable of. And even this task requires background knowledge: the machine needs to know the difference between like-sounding words such as 'there' and 'their', 'him' and 'hymn'. In short, the human brain understands language and how it works because of accumulated experience over the life of the individual. We are still a long way from grasping exactly how this happens, let alone how the neural connections are made that result in the sometimes odd and unpredictable conceptual links that result in 'great ideas'.

This book began with a look at some of our most basic abilities and ideas. Language is another such ability, some would say the most fundamental of all. It is fitting, then, that language should be so crucial to work on computers in general and on artificial intelligence in particular. For it points up how different we still are from machines, and what a long way our products have to go before they can have ideas like our own.

BIBLIOGRAPHY

This bibliography lists some of the works consulted in the preparation of this book and most useful to the general reader. Most contain further bibliographies which will lead the reader on to more specialist works.

General works

Birdsall, Derek and Cipolla, Carlo M., *The Technology of Man* (London, 1980)

Calder, Nigel and Newell, John, *Future Earth* (London, 1988)

Chant, Colin, *Science, Technology and Everyday Life 1870–1950* (London, 1989)

Clark, Ronald, *The Works of Man* (London, 1985)

Debeir, Jean-Claude, et al, *In the Servitude of Power: Energy and Civilization Through the Ages* (London and New Jersey, 1991)

de Bono, Edward (ed.) *Eureka!* (London,1974)

Macaulay, David, *The Way Things Work* (London, 1988)

Pacey, Arnold, *Technology in World Civilization* (Oxford, 1990)

Singer, Charles, et al, *A History of Technology,* 7 volumes (Oxford, 1978)

Temple, Robert, *The Genius of China* (London, 1991)

Architecture: pillar, vault, dome

Fletcher, B., *A History of Architecture* (London, 1975)

Harvey, J., *The Gothic World* (London, 1950)

Wheeler, M., *Roman Art and Architecture* (London, 1964)

Art

Bahn, P.G. and Vertut, J., *Images of the Ice Age* (London, 1988)

Laming-Emperaire, A., *Lascaux* (Harmondsworth, 1959)

Leroi-Gourhan, A., *The Dawn of European Art* (Cambridge, 1982)

Basic shelter; Stone and brick construction

Edwards, I.E.S., *The Pyramids of Egypt* (Harmondsworth, 1988)

Frankfort, H., *The Art and Architecture of the Ancient Orient* (Harmondsworth, 1970)

Beginnings of modern medicine

French, R., and Wear, A. (eds), *British Medicine in an Age of Reform* (London, 1991)

Shyrock, R.H., *The Development of Modern Medicine* (New York, 1947)

Boat, raft

Greenhill, B., *Archaeology of the Boat* (London, 1976)

Gilfillan, S.C., *Inventing the Ship* (Chicago, 1935)

McGrail, S., *Ancient Boats* (Aylesbury, 1983)

The book

De Hamel, C., *Scribes and Illuminators* (London, 1992)

Febvre, L. and Martin, H.-J., *The Coming of the Book* (London, 1984)

Gaskell, P., *A New Introduction to Bibliography* (Cambridge, 19XX)

Compass

Anderson, E.W., *Man the Navigator* (London, 1973)

Collinder, P., *A History of Marine Navigation* (London, 1954)

Computers

Babbage, H.P., *Babbage's Calculating Engines* (London, 1889)

Bowden, B.V., *Faster Than Thought* (London, 1955)

OECD, *Electronic Computers: Gaps in Technology* (Paris, 1969)

Concrete and steel construction

Giedion, S., *Space, Time and Architecture* (Cambridge, Mass., 1968)

Hitchcock, H.-R., *Architecture: Nineteenth and Twentieth Centuries* (Harmondsworth, 1970)

Richards, J.M., *An Introduction to Modern Architecture* (Harmondsworth, 1962)

Early agriculture, the plough

Barker, G., *Prehistoric Farming in Europe* (Cambridge, 1985)

Bender, B., *Farming in Prehistory: From Hunter-gatherer to Food Producer* (London, 1975)

Reed, C.A. (ed.), *Origins of Agriculture* (The Hague, 1977)

Rindos, D., *The Origins of Agriculture* (Orlando, 1984)

Early medicine

Baas, J.H., *Outlines of the History of Medicine* (Huntington, New York, 1971)

Marti-Ibanez, F., *A Pictorial History of Medicine* (London, 1971)

Electricity and the battery

Shearcroft, W.F.F., *The Story of Electricity* (London, 1925)

Tricker, R.A.R., *The Contributions of Faraday and Maxwell to Electrical Science* (Oxford, 1966)

Fire-making

Branigan, K. (ed.), *The Atlas of Archaeology* (London, 1982)

Flight

Mackworth-Praed, B., *Aviation: The Pioneer Years* (London, 1990)

Taylor, M.J.H. and Mondey, D., *Milestones of Flight* (London, 1983)

Hand tools

Grey, M., *Man the Tool-maker* (London, 1973)

Mercer, H.C., *Ancient Carpenters' Tools* (Doylestown, 1929)

Internal combustion engine

Hodges, D. and Wise, D.B., *The Story of the Car* (London, 1974)

Newcomb, T.P. and Spurr, R.A., *Technical History of the Motor Car* (Bristol, 1989)

Wise, D.B., *Illustrated History of Automobiles* (London, 1979)

Jet engine

Golley, J., *Whittle: The True Story* (Washington DC, 1987)

Whittle, F., *Jet: The Story of a Pioneer* (London, 1953)

Machine tools

Bradley, I., *A History of Machine Tools* (Hemel Hempstead, 1972)

Gilbert, K.R., *The Machine Tool Collection* (London, 1966)

Rolt, L.T.C., *A Short History of Machine Tools* (Cambridge, Mass., 1965)

Mass production

Herndon, B., *Ford* (London, 1970)

Metalworking

Aitchison, L., *A History of Metals* (London, 1960)

Friend, J.N., *Iron in Antiquity* (London, 1926)

Gale, W.K.V., *Iron and Steel* (Buxton, 1977)

Moving images

Bardèche, M. and Brasillach, R., *History of the Film* (London, no date)

Fielding, R. (ed.), *A Technological History of Motion Pictures and Television* (Berkeley and Los Angeles, 1967)

Geddes, K. and Bussey, G., *Television: The First Fifty Years* (Bradford, 1986)

Natural power

Beedell, S., *Windmills* (Newton Abbot, 1975)

Burton, A., *The Changing River* (London, 1982)

Reynolds, T.S., *Stronger Than a Hundred Men* (Baltimore, 1983)

Nuclear power

Hewlett, R.G. and Holl, J.M., *Atoms for Peace and War 1953–1961* (Berkeley and Los Angeles, 1989)

McKay, A., *The Making of the Atomic Age* (Oxford, 1984)

Optical inventions

Butler, S., *The Social History of the Microscope* (Whipple, 1986)

Ford, B., *The Revealing Lens* (London, 1973)

Turner G., L'E, *Essays on the History of the Microscope* (Oxford, 1980)

Pottery and glass

Douglas, R.W., and Frank, S., *A History of Glassmaking* (Henley on Thames, 1972)

Kingery, W.D. (ed.), *Ancient Technology to Modern Science* (Columbus, Ohio, 1984)

Plastics

Imperial Chemical Industries, *Landmarks of the Plastics Industry* (ICI, no date)

Katz, S., *Classic Plastics* (London, 1984)

Katz, S., *Early Plastics* (Aylesbury, 1986)

Remote communications

Hill, J., *Radio! Radio!* (London, 1986)

Kieve, J.L., *The Electric Telegraph* (Newton Abbot, 1973)

Young, P., *Person to Person: The International Impact of the Telephone* (Cambridge, 1991)

Roads and railways

Allen, G.F., *Railways Past, Present and Future* (London, 1982)

Ellis, C.H., *The Lore of the Train* (London, 1971)

Hindley, G., *A History of Roads* (London, 1971)

Nock, O.S., *Railways Then and Now* (London, 1975)

Rockets

Emme, E.E., *The History of Rocket Technology* (Detroit, 1964)

Haley, A.G., *Rocketry and Space Exploration* (New York, 1958)

Scientific agriculture

Fox, H.S.A. and Butlin, R.A., *Change in the Countryside* (London, 1979)

Franklin, T.B., *A History of Agriculture* (London, 1948)

Niall, I., *To Speed the Plough* (London, 1977)

Simple machines

Dircks, H., *Perpetuum Mobile* (London, 1861)

Strandh, Sigvard, *The History of the Machine* (London, 1989)

Sound reproduction

Jewell, B., *Veteran Talking Machines* (Speldhurst, 1977)

Steam engine

Riemsdijk, J.T. van and Brown, K., *The Pictorial History of Steam Power* (London, 1980)

Robinson, E., and Musson, A.E., *James Watt and the Steam Revolution* (London, 1969)

Tools

Hawkes, Jacquetta, *The Atlas of Early Man* (London, 1976)

Oakley, K.P., *Man the Tool-maker* (London, 1958)

Vaccination

Fisk, D., *Doctor Jenner of Berkeley* (London, 1959)

Wheel and wheeled vehicles

Piggott, S., *The Earliest Wheeled Transport* (London, 1985)

Science Museum, London, *Story of the Wheel* (London, no date)

Smith, D.J., *Horse-drawn Vehicles* (Cambridge, 1981)

Writing

Gaur, A., *A History of Writing* (London, 1984)

Hutchinson, J., *Letters* (New York, 1983)

INDEX